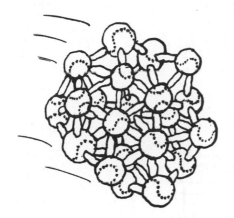

Cartoons by Tony Hall

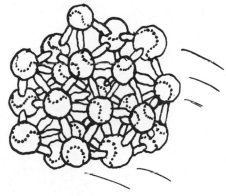

The Game of Science

Second Edition

Garvin McCain

The University of Texas
at Arlington

Erwin M. Segal

State University of New York
at Buffalo

Brooks / Cole Publishing Company
Monterey, California
A Division of Wadsworth Publishing
Company, Inc.

ISBN: 0–8185–0093–X

L.C. Catalog Card No: 73–75564

Printed in the United States of America

1 2 3 4 5 6 7 8 9 10—77 76 75 74 73

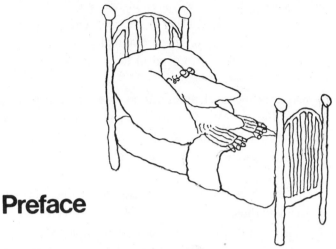

Preface

Science is a dominant theme in our culture. Since it touches almost every facet of our lives, the educated person needs at least some acquaintance with its structure and operation. He should also have an understanding of the subculture in which scientists live and the kinds of people they are. His understanding of the general characteristics of science as well as specific scientific concepts is made easier if he knows something about the things that excite and frustrate the scientist.

The second edition of this book, like the first, is for the intelligent student or layman whose acquaintance with science is superficial; for the person who has been presented with science as a musty storehouse of dried facts; for the person who sees the principal objective of science as the production of gadgets; and for the person who views the scientist as some sort of magician. The book may be used to supplement a course in any science, to accompany any course that attempts to give an understanding of the modern world, or—independently of any course—simply to provide a better understanding of science. We hope this book will lead the reader to a broader perspective regarding scientific attitudes and a more realistic view of what science is, who scientists are, and what they do. It will give him an awareness and understanding of the relationship between science and our culture and an appreciation of the roles science may play in our future. In addition, the reader may learn to appreciate the relationship between scientific views and some of the values and philosophies that are pervasive in our culture.

Throughout the book, we use "he" and "him" as pronouns; we mean them to include both sexes—as these words grammatically can—to avoid the awkward "he and she" and "him and her." We acknowledge that there are many excellent women scientists and expect a rapid change in our society so that soon no woman will be hindered in her academic and professional pursuits because of her sex.

We have tried to make the book entertaining as well as informative. Our approach is usually informal. We feel, as do many other scientists, that we shouldn't take ourselves too seriously. As the reader may observe, we see science as a delightful pastime rather than a grim and dreary way to earn a living.

A few of our comments obviously represent essentially unconfirmed personal opinions, particularly those concerning scientists themselves. Since the purpose of the book is to represent science as some scientists see it, we feel that these journalistic expressions are proper and fitting. Furthermore, since no one can write without making value judgments, rather than attempt to deceive the reader we make them blatantly and without apology.

Many activities and many disciplines fall within the broad realm of science. To keep our subject in focus we have emphasized one particular aspect of the scientific effort—one with a great deal of unity. That aspect is often called "basic" science: the acquisition of knowledge for its own sake. Unless otherwise qualified, basic science is implied whenever the word "science" is used. Some use the term "pure" science in the same sense we are using "basic." The choice of words is arbitrary.

Examples in the book were chosen to illustrate points rather than to provide comprehensive coverage. They have been drawn from friends and acquaintances as well as from many histories of science. We have deliberately chosen them from several different disciplines in order to show both similarities and differences within the scientific community. Although a substantial number were taken from some of the classical accomplishments of scientists, they do not represent any particular era; they simply illustrate the idea at hand.

When we present an example we generally identify the major persons connected with it in order to personify the examples and to acquaint the reader with a few classical scientists. The specific citations are not given in the text because we don't want to divert the reader's attention. However, a number of pertinent references are described briefly at the end of the book.

A book such as this necessarily contains a number of technical terms that may be defined somewhat differently by different scientists; in most instances we have used definitions that are acceptable to a large number of scientists. We generally define these terms explicitly when they are introduced.

It would be an understatement to say that a great many things have happened to the scientific world since we wrote the first edition of this book. One of the fundamental changes has been in the attitudes towards science of the public, the academic community, government, and indeed segments of the scientific community. These changes will be discussed at some length in Chapter 7. Attitudes are ordinarily reflected in some action and the attitudes toward science are no exception.

The authors have also changed over the last few years. One of us has changed his entire area of research primarily because the previous area had become somewhat boring and the new area appears both novel and exciting. The other has become very much interested in the problem of analyzing the process of "paradigm change," to be discussed in Chapter 4.

In spite of the scientific and personal changes of the past few years, the basic attitudes we expressed in the original preface have not changed greatly. This lack of change does not reflect simpleminded unawareness. Attitudes and analyses of science have been discussed for over 500 years. While both attitudes and analyses have been modified over the years, the changes have reflected changes within science rather than from external pressures.

The reader will note that in passing we have given the back of our hand to a number of groups. Most of these unkind cuts reflect commonly held attitudes among scientists of our acquaintance. Our own attitudes are not detached from all personal acquaintance with the groups involved. We have between us a number of years' experience in the military, business, administration, and applied science worlds. All of the maligned groups play an essential part of the modern scene. Their greatest fault is a narrow, self-satisfied view of themselves. Their resemblance to people, and scientists in particular, is rather striking.

Both of us owe much of the content of this book to what we have learned from teachers, colleagues, and the printed page. If this book has value, the credit should go to them. In particular we would like to thank Professors E. John Capaldi, Charles N. Cofer, Herbert Feigl, James J. Jenkins, and Paul E. Meehl for directly creating our interest in and contributing to our knowledge of the game of science. Without personal contact with such stimulating scholars we would never have begun, let alone finished, this book.

Certain scholars whom we have never met have helped make us aware of the real importance of the history of science in understanding the game. The most important of these scholars are Bertrand Russell, James Conant, and Thomas Kuhn. We would particularly like to single out Kuhn's essay *The Structure of Scientific Revolution* as the critical focus of our orientation.

We would like to thank the instructors who reviewed the first edition and offered helpful criticism and encouragement: D. Murray Alexander, De Anza College; Barry Anderson, Portland State College; Frank R. Blume, San Bernardino Valley College; Kenneth M. Coffelt, Tarrant County Junior College; Edward I. Fry, Southern Methodist University; Raymond E. Gotthold, Carlmont High School; Edward J. Kormondy, Oberlin College; Louis I. Kuslan, Southern Connecticut State College; C. W. Scharf, The University of Texas at Arlington; Paul E. Trejo, De Anza College; Anthony Trujillo, San Joaquin Delta College; Jules Wanderer, University of Colorado; and Joseph Zucca, Carlmont High School. We would also like to thank Kerry Barnes and Betty Jean McKnight, students at the University of Texas at Arlington, Brian McCain, at Schreiner Institute, and Janet L. Mistler, at the State University of New York at Buffalo, who read the book at the manuscript stage; and Gladys Wehner, Gloria Myers, and Karen Ensminger Hoyer, who somehow converted our illegible scrawls into readable type.

We would also like to thank the reviewers for the second edition: William T. Barry, Gonzaga University; James N. Bowen, University of Texas, Arlington; Paul T. Heyne, Southern Methodist University; James B. Hickman, West Virginia University; Louis E. Price, The University of New Mexico; Brent M. Rutherford, York University. Finally, we would like to thank Anne McCain Coffman and Carol Segal, who were incisive in their criticism and tolerant of our excesses.

We alone, of course, are responsible for any errors, misunderstandings, and faulty generalizations contained in the book. We can't blame the reviewers because at times we were obstinate and ignored their advice. As co-authors we share equally the responsibility for the book. The order of the listing of this responsibility was determined by an ace over a ten.

Erwin M. Segal   Garvin McCain

Contents

The Game of Science

Second Edition

Chapter 1
An Introduction
to the Game

The word "game" has been applied to a wide variety of human behaviors. There are psychoanalytic games and war games, games of chance and games of strategy, games for the young and for the old. There are many familiar games in which we are involved either as spectators or participants.

What is it that leads one to characterize an activity as a game? Wittgenstein used "game" as his example *par excellence* of a term for which all of its uses have a family resemblance but no specific common character. The wide varieties of games and their diverse attractions are shown by the following list.

1. Diversion (charades, girl watching, solitaire).
2. Amusement (catch, pin the tail on the donkey).
3. Competition (football, tennis, Monopoly).
4. Intellectual stimulation (chess, bridge, crossword puzzles).
5. Social interaction (post office, dancing, charades).
6. Completion (jigsaw puzzles, solitaire, crossword puzzles).
7. Chance (roulette, poker).
8. Strategy (chess, tic-tac-toe, war games).

9. Personal enhancement (keeping up with the Joneses, king of the mountain, chicken, follow the leader).
10. Emotional gratification (hunting, mountain climbing, sky diving, necking).
11. Curiosity (spelunking, touring, bird watching).

Different aspects of the scientific effort contain many of the attractions on this list. Because of the similarities between the attractions of science and those of a diversity of games, we can consider science a game. One qualification ought to be made, however; although we view science as a game, we note that it is a game played by professionals; and like all games played by professionals, at times it entails activities that are tedious rather than amusing.

Curiosity is probably the most important motivation a scientist has. Like a spelunker exploring a cave, the scientist eagerly seeks new information. Both the search for and the discovery of new information are motivated primarily by the scientist's curiosity. Because of this, the scientist designs experiments, collects data, and then passes on to new vistas.

The world of the scientist is uncertain and incomplete. He is faced with a myriad of bits and pieces of data and ideas. One of his principal goals is to impose order on chaos. The major difference between putting together a jigsaw puzzle and organizing the elements of a science is that the scientist's puzzle is never completed.

Other aspects of games are present in scientific activity. Social interactions among scientists quite often lead to the refinement of scientific ideas. Many discoveries come to the prepared scientist by chance. Despite this, most scientific advances come from the strategic use of the scientist's weapons, and many scientists compete with each other in an attempt to get their ideas accepted by others.

One of the strong motives for many scientists that is related to competition and personal enhancement is the desire for recognition by others. One form this ambition takes is seeking to win the Nobel Prize. It's clear, for example, that James Watson was competing with Linus Pauling to understand DNA first and win the award. This is not so different from athletes going to Olympic games to win gold medals. One last motive that readers have suggested is not usually found in games—improving the world. Although the applied scientist is likely to consider this noble but distant goal, it is quite often subordinated to the everyday goals mentioned above. In fact, some scientists were actually disappointed when Salk and Sabin discovered a vaccine against polio because they had hoped for the glory of discovery for themselves.

Scientists working on a problem become very enthusiastic and emotionally involved. To understand a colleague's idea, to generate an idea, and to speculate on possible alternatives are all rich sources of intellectual stimulation. Rather than trivial, this stimulation is one of the most pervasive sources of pleasure for scientists. The game of science can be all absorbing; it can define a world into which the scientist can escape, body and soul. He can forget the dull, humdrum everyday routine where he, like the rest of us, spends a good deal of his time. The passion of intellectual challenge can compete successfully with any of man's other passions. Intellect cannot be separated from emotion, and progress toward solving scientific goals can arouse one to a feverish pitch and carry him away. It is interesting that although he may feel physically drained after such an experience, total involvement can be sustained for long periods of time.

This book describes the game of science, and it is organized around that theme. We have just discussed some of the reasons why the game is played. In Chapter 2, we emphasize what kind of a game science is and how both spectators and participants view the game. Chapters 3, 4, and 5 spell out how the game is played, its rules, and the varied activities of its participating scientists. Chapter 6 describes who the participants are, what motivates them, some of their values, what they believe, and how they become scientists. Chapters 7, 8, and 9 discuss the results of the game—what it does for and to society.

Some readers of the first edition of our book objected to characterizing science as a game. Any endeavor that leads to bombs big enough to kill 200,000 people at one time is too serious to be called a game. We do not deny the seriousness of science—it can be an extremely serious endeavor that has widespread, lasting, and even devastating effects on man and his environment. However, one of our biases is that human activities—especially those to which people are committed—have to be justified as ends in themselves as well as means toward other ends, since a lifetime of work must be fulfilling in itself and not simply have worthwhile ends. We note the similarities of scientists' motives for doing what they do to those of individuals involved in less serious games. In fact, we believe the game analogy to apply to all worthwhile human enterprise. We just hope that those who play serious games play them well.

To play or to observe the game of science requires some effort and a skill that takes time to perfect. As with many other games, beginning play may take one of two forms, whether one is a player or observer. The going might be tough and seem not worth the effort, or one might play superficially and feel that the game is trivially easy and not worth learning in detail. We believe that neither need be the case with this book. There have been many occasions when one of the authors has read an article or a book and found it to be trivial. After being called a blockhead (or worse) by a colleague (at times the other author) rereading the article would reveal it to be very valuable. On other occasions we have read an article and found it extremely hard to comprehend. In many of these cases rereading has simplified the article immensely. We do not understand how this process works, but we profit from our experiences. We believe science is an engaging game which can be rewarding to both the players and the observers when they understand what is going on. We think this book has captured some of that drama. If it seems trivial, we hope you look again to see if you missed something. If it seems too tough, remember that the results are worth a little effort. One is not educated if he does not understand something about the game of science. We hope to educate as well as entertain. If you have not yet done so, please read the preface before reading the rest of the book so that you are aware of the framework within which we wrote the book.

We do have a profound admiration for scientific achievements, and when we mention a conflict between science and external forces our prejudice is typically on the side of science. This does not mean, however, that we are unaware of the human foibles and failures within science.

Science is not infallible. Until 1962, chemists were convinced (and the texts said so) that the "noble" gases (neon, argon, krypton) would not form compounds. Accepted chemical theories recognized this "fact." Unfortunately for them, 31-year-old chemist Neil Bartlett was able to produce compounds from these gases.

Those who have read James Watson's *The Double Helix* or C. P. Snow's novel *The Search* (we recommend both) are aware that scientists have at least their share of human vanity, greed, lust, sloth, and indecisiveness.

Scientists have been remarkably unsuccessful in helping nonscientists to understand science. Even captive audiences in educational systems generally depart as scientifically innocent as on their day of arrival.

Lest the reader believe "true" science never takes a pratfall, he is invited to read the great American chemist Irving Langmuir's paper, "Pathological Science," describing the detailed work of some eminent and not-so-eminent scientists (mostly physicists) who found things that just weren't there. Yet hundreds of papers were published supporting these illusory results.

These statements are not meant to indict science or scientists but rather to recognize human limitations.

No one can write a book without making some value judgments. We affirm the view that value judgments are both proper and necessary. Clearly we made the judgment that science was worth writing about. This is no apology, merely a warning. We attempt to identify the more clearly unsupported personal views as they occur.

One final note: at the end of the book is a bibliography that includes a short description of each book. We hope this will guide you to some interesting and pertinent reading. Also useful for term papers.

Chapter 2
Attitudes and
Science

Do you believe in witches? Some people do. There has been a rebirth of interest in the supernatural during the last few years. For 100 years or more such beliefs were held only by a very few people in western culture. Why do we or don't we believe? Our belief system depends primarily on our general attitude toward knowledge. Our society and that of our ancestors contained roughly equal information about supernatural phenomena. The scientific attitudes prevalent in our society today are chiefly responsible for our disbelief in the supernatural.

Most people don't actually have scientific attitudes, but our cultural beliefs are influenced by the fact that a small but important minority does have them. An interesting aspect of the scientific attitude is that, although it affects beliefs, specific beliefs are not its most important product. According to Bertrand Russell, ". . . it is not *what* the man of science believes that distinguishes him, but *how* and *why* he believes it. His beliefs are tentative, not dogmatic; they are based on evidence, not on authority."

An individual's particular discipline (field of science) is not the key to his possessing or understanding scientific attitudes. There are those trained in highly theoretical and advanced sciences, such as physics or astronomy, who not only don't know how to play the game of science—they don't even know the game is going on. On the other hand, in less precise sciences such as sociology or economics, where it is more difficult to maintain scientific attitudes, there are many individuals who fulfill all of Russell's criteria. Attitude is distinctly more important as a criterion of science than is the sheer amount of solid data available or the degree of development of a particular science.

The Importance of Attitudes

At various places and times in the human struggle, there has been a variety of dominant attitudes or modes of thought. Nationalism, theology, the exquisite glories of war, business, and racism have each at one time or another functioned as the primary yardstick for human belief systems. Consider one of these yardsticks during the sixteenth century—theology. At that time theology clearly dominated the thought of Western Europe. Of course not everyone was a theologian. For example, only a minute portion of the populace had even a foggy understanding of the ontological argument (an argument for the existence of God). But the dominant mode of thinking involved theology at some point. It is equally evident today that, though science plays a dominant role in our society, only a small minority has any serious understanding of it. We are all aware of the attempts to link science with such fascinating subjects as deodorants, gasoline additives, and false teeth. Whether or not these links are legitimate, they exemplify the advertisers' confidence that "science" has sufficient glamour and acceptance to be worth a few dollars' extra profit. Implied is the further assumption that the viewing, reading, or listening public does not understand science, so that an oblique reference to "science" will reduce them to helpless acquiescence. Whatever the effectiveness of this approach, it does reflect an assessment of popular views. Many people seem to believe that production of better monsters, tranquilizers, satellites, bunion pads, or more humane bombs is the aim of science, and that the method by which these items are developed is through a collection and cataloging of facts. Such "practical" items may be necessary for late, late shows, employment in aerospace industries, comfort for aching feet, and waging wars, but their relation to science is rather remote. The popular conception of science thus emphasizes the production of gadgets, mysterious midnight work in the laboratory, and data collection and storage. Our aim is to suggest a somewhat larger perspective regarding scientific attitudes and to present a more realistic view of what science is and what scientists do.

It needs to be emphasized that although science affects belief, it is primarily an attitude toward problem solution; as such, it has implications far beyond the immediate data. For example, when Copernicus and Galileo shattered the earth's pretensions as the center of the physical universe, the impact was much greater than the astronomical consequences of accepting the sun as the center of the universe. Presumably, the Copernican concept of the universe did not change the productivity of the fields, turn wine to vinegar, or render martial and marital games less fascinating. And yet, when a long established and firmly held belief is shattered, life can never be the same. At least as menacing but more formless was the thought that if one fundamental truth had to be discarded, then, like the hole in the dike, a flood of unknown and frightening proportions might follow.

Science is very much like other human endeavors in that attitudes are of extreme importance and reach far beyond the immediate effect. Consider the doctrine of the "divine right of kings." At one time this idea was widely accepted, but by the latter 1600s it was seriously challenged or rejected through most of Europe. Although the initial change of attitude involved primarily the monarch, it foreshadowed curtailment of the privileges of the aristocracy. Later consequences of this shift in attitudes included the doctrine of equality of men and the questioning of hereditary property rights. It would be absurd to assume that all of these later developments in social structure resulted solely from the overthrow of the divine-right doctrine, but the change in attitude accompanying this overthrow did play a substantial part in their development. Today it is extremely difficult to understand, much less re-create, the attitudes and implications that accompanied acceptance of the divine-right doctrine.

In summary, when we change the way we view any important aspect of our world, our attitudes toward other aspects of the world also change, because attitudes are generally more important than specifics.

The Nonviolent Nature of Science

Although science is often revolutionary and the *products* of science may well terminate life on this planet, the *problems* of science are not soluble by violent means. This nonviolent character is not due to any angelic aspects or overwhelming virtues of scientists. Even if a violent outbreak were to occur and one scientist physically clobber another, the scientific problem would not be resolved. The simple fact is that scientific problems do not lend themselves readily to violent solutions. A nationalist can make his point and achieve territorial aspirations with a bayonet. An economic philosophy can gain a place in the sun following a violent revolution, as Lenin and Mao have demonstrated. "Conversion by the sword" has had lasting effects in many parts of the world. But scientific struggles have been relatively bloodless. Galileo and his small group of followers won their struggle against the church, the strongest power of their time. There is no record that he or any of his followers employed violence in the contest.

Stalin used violence, threats, and bribes to impose political standards on genetic theories in supporting Lysenkoism, a theory holding that acquired traits are hereditarily. transmitted. His success was fleeting and disastrous. Why? The attempted application of unconfirmed theory brought Soviet genetic research to a virtual standstill and created difficulties in agricultural development. Scientific theories must at some point be anchored in observations. Try as you may, you can't force observable data to change; they are stubborn and eternally patient. If observers go to the stake or to Siberia, there will be others to observe; observers may remain silent, but the data become like heavily starched underwear—concealed from others but difficult to ignore.

Scientific controversies are in part settled by appeals to evidence. This evidence includes observations at some point. There are no time limits on the accumulation of evidence, since controversies can be reopened or continued indefinitely. Typically, settlement of a controversy takes place over a period of time; personal prides and prejudices are slowly dissipated and are not likely to lead to violence.

Controversies are generally resolved when a scientist submits his findings and speculations to a jury of his peers. Only their judgments count. Although these peers make mistakes, they allow continuous and unlimited appeals of adverse judgments. Like an uncertain shopper, scientists are rarely forced to make an immediate conclusion, and when a transaction has been completed the "merchandise" may still be returned. Such a system can be magnificently frustrating, but still, controversies are ordinarily resolved without violence.

Hostility toward Science

Despite its nonviolent nature, its glamour, and its importance—or perhaps because of these qualities—science has engendered underlying streaks of hostility in practically all levels of our society: fundamentalist religious groups claim that science inveighs on their province and would destroy all religion; the humanist charges that science dehumanizes man and reduces him to an automaton with no special quality or purpose; the conservationist accuses science of turning our environment into a shambles with no forests to visit, no animals to see, no challenges to face, no water to drink, no air to breathe; the New Left states that science threatens their lives with the bomb, deliberately destroys cultures to save them, denies people their freedom, and doesn't allow normal human existence to proceed; and the politician sneers that science never accomplishes anything and is a great waste of money.

Most of the hostility toward science can probably be divided into two types. First, people oppose new ideas that challenge their beliefs. Second, some people oppose the application of the conclusions and technology of scientists to real-life situations. There is, in addition, hostility toward science from the scientific community. In many cases, the scientific establishment inhibits and attacks both new ideas and the scientists who present them. The reciprocal of this hostility is also present. Many scientists are hostile toward what they deem the rigidity and self-serving nature of much of the scientific community with its implied absolute truth and morality. These scientists are also hostile toward many real-life decisions made in the name of science. In other words, there is strong hostility among scientists as well as between the rest of society and scientists. The two types of hostility toward science operate in microcosm within science.

The first type of hostility has had a long and sometimes bloody history, with scientists providing the blood. However, many of today's scientists have become free to present their opinions. Physics, in particular, has become relatively safe, primarily because it is now so complex and abstract that only a very foolish layman can even pretend comprehension. Besides, there has been so much technological confirmation of theories of physics in our everyday life. We have electricity, automobiles, telephones, and refrigerators. We can read about and see such delights as spaceships to the moon and hydrogen bombs. Who would challenge a science that has made it possible for Lawrence Welk to appear in our living rooms every week?

At the other end of the spectrum, however, are sciences that are often not recognized as sufficiently abstract to deter the layman. In addition, these sciences tend to be closely related to some vested interest. At present, sciences such as sociology, psychology, and economics pose the same sort of threat to parts of our society as Darwin's biology and Galileo's physics and astronomy did earlier. Many nonscientists "know" the answers to problems studied by social scientists through their own experiments, observations, introspections, insights, and discussions. We all have belief systems that require certain statements to be true, and we will not accept denial of these truths from anyone. One's way of life—whether Eskimo or Texan—is usually endowed with unique virtues. Even views that seem impossible to test can be held with a vigor out of all proportion to their importance. How many violent discussions originate with a comparison of sports teams or stars, films, or actors? Such discussions generally contain a minimum of evidence and a maximum of feeling. The whole point is that we become emotionally involved with our beliefs, and woe to him who challenges them, no matter what his argument or evidence. Thus the social sciences, which often challenge our personal beliefs, are applauded when they agree with us and scorned when they disagree.

The social scientist's role is further complicated because each scientist has his own personal belief system. Ideally, a belief system should not influence scientific judgment, but the ideal is not always achieved in practice.

Whether currently articulated specific principles from the social sciences prove useful in the long run remains to be seen. However, within each of these disciplines, at least some people understand and play the game of science. In the future, many of their revolutionary hypotheses will be blandly accepted with the same lack of emotion that now accompanies a statement that the earth circles the sun.

We mentioned psychology as a social science—actually, psychology is not a discipline, it's a collection. Orientations in psychology range from social (social psychologists) to mathematical (statisticians and math modelers) to zoological and biological (comparative psychologists) to anatomical, physiological, and biochemical (physiological and pharmacological psychologists). This description only scratches the surface. When we refer to psychology as a social or biological science, we are referring to one of its subdivisions.

Our modern society is somewhat different from the societies of Darwin and Galileo. Today, even in controversial areas such as the social sciences, one can speak freely. Of course, free speech may still carry fringe benefits such as denunciation in the press, a sudden change of job locale, or censure by public or private bodies. For example, B. F. Skinner, an outstanding psychologist, recently published *Beyond Freedom and Dignity*, in which he concluded that the majority of man's behavior is learned. He also proposed that, since important ethical decisions are learned on some arbitrary basis, we should deliberately shape the child's attitudes toward ends that are beneficial to society. At least one Congressman attacked Skinner's book and denounced the use of federal funds for research that led to such conclusions. In a somewhat similar situation, the U.S. Senate passed a resolution condemning the Report of the Commission on Obscenity and Pornography (the report was based on findings by a large number of reputable scientists and scholars) without reading it—in fact, even before the report was printed! Forty-seven years after the Scopes "monkey trial," teachers are still being attacked for teaching evolution in the science classroom. Clearly there are still those ready to condemn an idea, a conclusion, or a man without waiting to be confused by evidence or logic.

The second major source of hostility toward science is a much more modern type. This attack claims that science is misused in one endeavor or another. Here we have a direct conflict between ideology and the development or application of scientific knowledge. Many people in our society simply object to many of the domains to which scientific knowledge (or supposedly scientific knowledge) is being applied. Many students of science seriously object to scientific studies being sponsored by the Department of Defense (DoD). They envision the purpose of DoD as waging war and feel that DoD is using scientific knowledge to enable man to kill man more efficiently. Although much of DoD's research money is not used in any direct way to wage war and much of it is for basic research, to many it symbolizes destruction rather than construction; so they oppose both the DoD and anything that it supports. No one can deny that much DoD research money is being used to develop better delivery systems for weapons, better weapons, better evaluation of military personnel, or better spying systems; so there is no clear, simple relation between DoD and publicly available scientific research.

Many funds have been going to very expensive National Aeronautics and Space Administration investigations. The major stated justification for these expenses has been scientific exploration, although their military value has not been overlooked. Many people have noticed all the problems in our society that cannot get funds for investigation and hold science as well as the industrial and governmental establishment responsible. There is much discussion about national priorities; a large part of the scientific community has supported the priority system that we tacitly follow.

Certain members of the scientific community have been involved in application of scientific theory to power politics. Whether the domain is the war in Viet Nam or electing someone to the Senate or presidency, many people feel that the public is being manipulated for the goals of a few. We need not discuss whether it is "good" science or "bad" science that leads to these manipulations, but it is clear that many people hold science responsible. Many scientists do, in fact, participate in the decision processes of war and politics, so those who object to the decisions easily extend their hostility to science.

In addition to theoretical and practical disagreements with scientists, hostility may result because of certain personal characteristics they are seen to have. Scientists tend to qualify their answers; for someone accustomed to a "yes" or "no" answer, this tendency can be annoying. (We'll examine the reasons for these qualified answers later.) Many scientists even refuse to discuss their findings with representatives of the popular media. The public ought to be informed, but unfortunately a scientist's statements and the final news-media presentations are often about as closely related as fifth cousins twice removed. The blame does not rest solely on the media. If the scientist simplifies, he aids the distortion and appears simpleminded to his peers. If he explains concepts and data in detail, he puts the reporter through a Ph.D. program during the interview. The problem is very real. If this book is successful in giving the reader some insight into this problem, it will have achieved one of its purposes.

Scientists are often portrayed as intellectually arrogant and unwilling to listen to or consider the views of others. There is more than a germ of truth in this sort of accusation. However, they tend to be arrogant primarily in their own fields. Such attitudes are not confined to scientists. They exist in most if not all highly professionalized groups. For example, the major-league baseball player may refer to less accomplished players as "bush leaguers" or "sand-lotters"; these titles reflect neither overwhelming humility nor a high regard for the less gifted. Discuss your favorite home remedy with an M.D.; your suggestions for running a business with an executive; your ideas of art with a professional artist; philosophy with a philosopher. Then judge whether the scientist is alone in his arrogance.

Scientists are alleged to be irreligious, unconventional, and of doubtful loyalty. The charge of irreligion could be more accurately phrased. If the charge is changed to a lack of identification with religion, evidence shows that the description is accurate for a great portion of the active scientific community. Many scientists find it difficult to reconcile scientific attitudes and religious feelings and beliefs. The result is that a great portion of the scientific community simply ignores organized religion.

At one time many scientists were actively antireligious because of the open conflicts between religion and science. Today, religious groups no longer have the power to suppress scientific advancement. If they reject certain scientific theories or discoveries, their influence is confined largely to the sympathetic audiences of their own members, and scientists can usually ignore them. Of course not all scientists are uninterested in religion. Repeatedly, studies indicate that although active scientists have little or no interest in organized religion, inactive scientists tend to have at least a moderate interest in the subject.

The charge of unconventionality of scientists depends very much on one's definition of convention. Within their own subculture, scientists tend to be rather conventional—although what is accepted as conventional in the scientific subculture still allows its members substantial latitude. The "unworldly" attitudes of scientists toward wealth is an example. As contrasted with the prestige of a businessman among his peers, a scientist's prestige in the scientific community bears little or no relationship to his wealth. In this respect the scientist is no different from the businessman; each conforms to the values of his own group.

However, conforming to the scientific subculture is not the whole story. In addition to having different values, the scientific community has a substantial number of members who may not fit any identifiable set of conventions. The demands of the scientific game are such that high levels of independence and intelligence are required of many participants. Even in these cases, however, the individuals may be not so much nonconformist as simply unconcerned with a particular set of values.

The charge regarding loyalty is difficult to answer simply. Science is not limited by international boundaries or institutional alliances. From the very early days, men of science have maintained contact through correspondence, conventions, and laboratory visits. A scientific triumph is part of the common enterprise, without regard to its place of birth. Science is, and must be, a public endeavor. A scientist's whole training and way of operating cries out against secrecy. His profession requires information in order to progress. He tends to be loyal to his profession and his colleagues. One way of looking at the problem of loyalty is to realize that all individuals have multiple and often conflicting loyalties. A husband and father's loyalty to his family, a religious man's loyalty to his particular god, or a scientist's loyalty to his search for understanding pose different and difficult conflicts with other loyalties. Every person has to resolve the problem of his own conflicting loyalties. It is not implied that scientists have handled themselves badly in this field. Rather, the wonder is that with all their possible conflicting loyalties they have handled themselves so well.

In short, scientists are neither as appealing as teddy bears nor as offensive as skunks. For better or worse they are products of their subculture and the types of personalities it attracts. It is interesting to note that the relation of scientists to their society is not limited by national boundaries; Soviet scientists seem to have the same abrasive qualities within their society that our local products have in ours.

Relevance

Many people see much, if not all, of science as irrelevant. Although this might not generate hostility, it does turn people off. This is an important problem because the future of science and the development of our culture depends to a great extent on the young men and women who become scientists. Also, in a culture that is to a great extent dependent on science, all citizens should have some appreciation of science, which is impossible with a turned-off attitude.

The concept of relevance has two aspects. Many people believe that scientists should work directly on practical problems. They feel that basic research, which has to do primarily with understanding and theoretical development rather than solving immediate problems, is thus irrelevant. This is not the most important definition of irrelevance, but it should be briefly considered.

Research on the sex life of insects has been used to aid in the elimination of pests. Trying to understand lightning led to the harnessing of electrical energy for all of its myriad uses; investigating filterable life forms led to the polio vaccines; thinking about what the world would look like if one were to ride on electromagnetic waves led in part to nuclear energy; and having animals press bars in boxes led to helping the mentally retarded. It isn't that the applications of science are all accidental, but that topics that seem to be eons apart may be related to each other; at times one finds the solution to one problem by working on another that is related to it. In addition, if a scientist has a practical problem to solve, he may find that it cannot be solved directly, but that he has to solve corollary problems first. The problems associated with immune mechanisms and rejection of foreign bodies have to be solved before organ transplants can be generally successful. In fact, much of the research in social sciences—which seems very theoretical and irrelevant to real issues—started with the attempt to solve practical problems.

It is not the lack of practical applications that usually turns people off from science, however. Relevance is more an emotional problem. Some things seem dull, uninteresting, and intuitively wrong—irrelevant to what one thinks and believes. This is the irrelevance of science that really matters.

Although we are not sure why people get turned off, some reasons have been proposed. C. P. Snow has argued that two alternative cultures compete in the intellectual world—the scientific and the literary. There is minimal communication between these cultures, so if the student begins exploring one, he tends to reject the other. One of the major differences between these cultures is that the members of the scientific culture tend to be much more optimistic about what can be done for the fate of man, and those in the literary culture feel that nothing can be done to change man's fate, and therefore he must live each "existential moment" for itself. There has been a great deal of pessimism in the United States since the assassination of President Kennedy. This pessimism has been supported by other assassinations, the continuing war in Viet Nam, the increased awareness of pollution, poverty, and other problems. Considering our culture's failings, it is not surprising that many have turned away from the optimism of science and toward the attempt to accentuate feelings and the existential movement. Science seems irrelevant because the feeling it conveys is counter to the feelings of a growing number of people. (We hope this depression will be short-lived, that progress will be made, and that some optimism will return.)

Another reason that many people tend to be turned off from science is that they may read so-called relevant literature. For example, there has been a vast amount of publicity on marijuana use. Millions of Americans have smoked pot. Both those who have and those who haven't smoked may be interested in the subject, and a lot of scientific research has been done. What do these people interested in a relevant topic find out? They may find that naïve users of marijuana may have less coordination of small muscles, dryness of the mouth, and a slight increase of heartbeat rate. About 90% of marijuana research is biological and biochemical in nature, and there is no description of the experience of a high itself in its different forms. Readers infer that their psychological reactions are irrelevant to science and vice versa. However, we believe that scientific methods can be successfully applied to experiential domains, although very little has been done so far.

Finally, regardless of its subject matter, scientific prose can turn anyone off. Most of it is totally impersonal and unemotional. The dynamic, emotional, competitive, exciting reality of scientific discovery is presented in the passive voice, so we read that experiments are run and theories are explained, but the scientist himself never does anything. The problem of communication from the scientists to the layman is similar to that of any expert to one outside the field. The dreary but critical details of procedures, the complexity of interpretation, the difficulty of terminology and many other factors make the attempt to communicate on a professional level in layman's terminology a nightmare.

There are inevitably points on which co-authors do not agree. The problem of communicating scientific work is one such problem. We both agree that clarity, simplicity, and readability are goals to be fervently sought. We disagree about the extent to which it is possible to use language understandable to the layman in such communication. One of us feels that much more could be clarified if a more readable style were used; the other stresses that there are limits to what a layman can understand no matter how clear the language.

Bertrand Russell preferred plain English; he explained how he got away with using it:

> I am allowed to use plain English because everybody knows that I could use mathematical logic if I chose. Take the statement: "Some people marry their deceased wives' sisters." I can express this in language which only becomes intelligible after years of study; and this gives me freedom. I suggest to young professors that their first work should be written in a jargon only to be understood by the erudite few. With that behind them, they can ever after say what they have to say in a language "understanded of the people."

On the other hand, Russell and Alfred North Whitehead did write *Principia Mathematica*, which revolutionized mathematics, although it has been stated that very few people in the world—perhaps six?—have read the entire book.

One possible solution is separate literatures for the expert and the layman. The principal objective would be to give the layman an understanding of what Conant calls the "tactics and strategy of sciences." Summaries of recent findings in science and accounts of the actual processes and human beings involved should help.

The question remains: is it possible for scientists, mathematicians, engineers, or other such specialists to write in such a way as to be bright, sparkling, and understandable to the layman and at the same time convey adequate information to other specialists?

As a final thought connected with relevance, we need to consider the possibility that scientists can influence social processes in a society such as ours. Scientists' primary weapons are not their numbers or financial power, but their knowledge and ability to influence, through rational presentation, political and economic processes. Sounds simple and workable—or is it? Take the example of air pollution. Automobile exhaust emission is related to the amount of fuel burned in our engine. Suppose scientists proposed a limit on the size and fuel consumption of automobile engines. Would they find opposition to what appears at least a debatably rational suggestion? Consider only a few sources: dowagers who consider block-long hearses as the only dignified way to travel, teen-agers with four on the floor, hot rodders, oil companies, service stations, tire manufacturers, automobile manufacturers, highway construction companies, steel companies, etc, etc, etc. In time, possibly something could be done. Some have suggested that scientists band together and refuse to work on harmful projects. What real effect would a strike have in the case of automobiles?

The Assumption of Simplemindedness or Omniscience of Scientists

The title of this section may seem somewhat strange, but it does represent frequently encountered attitudes. On the one hand, people often discount or scoff at scientists when a theory violates common sense or threatens some cherished belief. On the other hand, the atom bomb, polio vaccine, and rocketry have surrounded scientists with an aura like that around ancient magicians. The irony of this situation is that practical products are viewed with awe while theoretical structures may be considered hopelessly moonstruck. Interestingly enough, scientists tend to evaluate these accomplishments in the opposite order: they value theoretical contributions far above the practical products of those theories.

One of the best examples of the layman's attitude is the man who announced in a confident bray that he could prove Einstein was wrong when he said the universe is limited simply by asking one question: "What is on the other side of it?" The simple question designed to put a scientist in his place is not exactly a new or surprising idea. The great poet Goethe attempted to refute Newton's concept of the composition of white light by looking through a prism at a white object. He could see by simple observation that white light was not composed of many different colors, it was just plain white.

In some ways, scientists can sympathetically understand and even agree with laymen: The earth is indeed flat, and the sun does go around it; mash a toe and the pain is immediate; a chair is a solid object, not mostly empty space (easily confirmed when it meets a shin in the dark). For most everyday purposes these answers represent perfectly reasonable ways to view the world, and should you behave as if these statements were correct you would be unlikely to come to serious harm.

The scientist's way of looking at the world is also reasonable and useful. An important reason for the layman's assumption that the scientist is simple is that a scientist often deals with concepts and data that go beyond immediate sensation. Remember, the scientist has a large store of observations and concepts from many sources, and they need to be integrated in some way. For example, the scientist knows that the speed of nerve impulses varies from about 0.7 meters to about 120 meters per second, depending on the type of nerve fiber. Although pain seems instantaneous, substantial evidence indicates that it is not. This evidence has been gathered by surgeons using an electrical stimulus, a microsecond timer, and electronic recorders. This work is presented in most good physiology texts. The chair example mentioned above illustrates the problem of integrating both observations and concepts. With the proper instruments we can observe that the wood in the chair has a definite cellular structure that is not at all apparent to the immediate senses. Can we abandon our notions of atoms and molecules derived from, and important in, other contexts just to render our chair as solid in all ways as it feels to the shin?

The object of these examples is to emphasize that the scientist, in his world, playing his game, has different problems from those of the layman. These problems require what may seem rather strange answers. The scientist is not unaware of the raw appearance of things. Rather, he has to make his answers consistent with a wide range of data and concepts as well as raw appearance. He may be confused or wrong concerning a particular concept, but the reasons are not likely to be apparent to the untrained eye.

"Smog, cancer, intergalactic travel, juvenile delinquency, birth defects, schizophrenia, universal education—science can solve any problem if scientists will only work on it." This is a popular version of the omniscient view of science. Scientists themselves are generally much more cautious; they doubt their abilities more, and they place great emphasis on the difficulties of even partial solutions of these practical problems. In many instances scientists do not even know how best to try to solve a practical problem. Unfortunately there are many problems for which science has no answers at present and others for which it may never provide answers. Another aspect of solving practical problems is that the technological aspects are not enough. On many an occasion the technical problems have been solved but the distributional problems remain. Even if a problem is solvable, it is far from certain that the solution will be put to use where it would do the most good.

President Truman once made an observation that might well illustrate a reasonable expectation of scientists. Carved on the tombstone of a Midwesterner who died in the 1800s were the words "He done his damndest." Our problem is that sometimes the scientist's damndest is not enough when it comes to solving important problems.

Misunderstandings about Science

There are a number of common misunderstandings about science. The following statements illustrate some of them. Each of these misunderstandings is discussed briefly.

1. The accumulation of facts or data is the primary goal of science.
2. Some sciences can be described as exact.
3. Science is deficient because it cannot give any ultimate explanation of natural processes.
4. Scientists distort reality and cannot do justice to the fullness of experience.
5. Science is concerned primarily with man's practical and social needs.

1. *The accumulation of facts or data is the primary goal of science.* An important part of the game is collecting data, but mere accumulation of facts does not necessarily give birth to science. All human groups collect data, but not all humans are scientists. Consider the following vignette. As in a dream, transport yourself to a gentle isle of the South Seas. Here frolic the happy children of nature, their golden skins glowing amid glittering sands, azure seas, and lush foliage. Dominating this spectacular beauty is a mountain from which a pale trickle of smoke arises. What do the people know about this object in their midst? Quite a bit! They have amassed a wide range of data. They recognize the rumbles that precede a shower of ashes; know that the refuse of an earlier eruption is extremely fertile; know that ashes and lava are hot and have a distinctive odor. Legends of the village tell of fireworks bursting high above the mountain.

We could continue to catalog their knowledge indefinitely, but a simple recital of data has not led to any "scientific" understanding. The missing ingredient is a set of general principles. The inhabitants' accumulation of data is extensive, but aside from the classic line "The gods are angry tonight," they make no sense of the welter of events.

To summarize the point of this illustration, data have been accumulated in all cultures, but science is recent and flourishes in only a few. The quality and character of explanatory or organizing concepts and the relation of concept to data are critical additional ingredients in the game of science. The concept of "angry gods" does not lead to any reliable prediction of future events; nor does it integrate specific observations into an organized pattern. In addition, "angry gods" will not hold still for testing.

2. *Some sciences can be described as exact.* There are no "exact" sciences! And there is no prospect that there will be exact sciences in the future! It may occur to you that mathematics is "exact." Whether it is or not is beside the point, since mathematics, although vital to science, is not itself a science, as we define the term. Science must be related to observations, and mathematics is not derived from nor dependent upon observations—it is simply based on the logical consequences of a set of postulates. Some people call observationally based sciences "natural" sciences and mathematics and logic "formal" sciences.

Statements made in natural science have only probabilistic, not exact, confirmation. Take as an example a long established and repeatedly confirmed scientific law. Consider the distance (S) traveled by a falling body in its first two seconds of fall. The established law is presented as $S = \frac{1}{2}gt^2$ (g is a "constant" that varies with location and altitude; t is time in seconds). Actual measurement of distances (taken at St. Louis, Missouri), however, would generate a distribution of values similar to that in Figure 1. In other words, repeated measures under the same conditions (as nearly identical as humanly possible) give different answers. The data conform to laws of probability which imply that one gets a particular range of results a certain percentage of the time. There is no way we can determine exactly how far the body falls in a given amount of time under any specific condition. What we can do is state with a high degree of assurance that in this case 95% of the measures will fall within 0.0001 meters from the average.

In addition to their probabilistic nature, scientific statements are incomplete. Although it is generally unspoken, a qualification is implied in any scientific statement· "Based on the evidence available to me, interpreted to the best of my ability, I believe that. . . ." In other words, a scientific statement is inevitably incomplete in evidence and interpretation. "Evi-

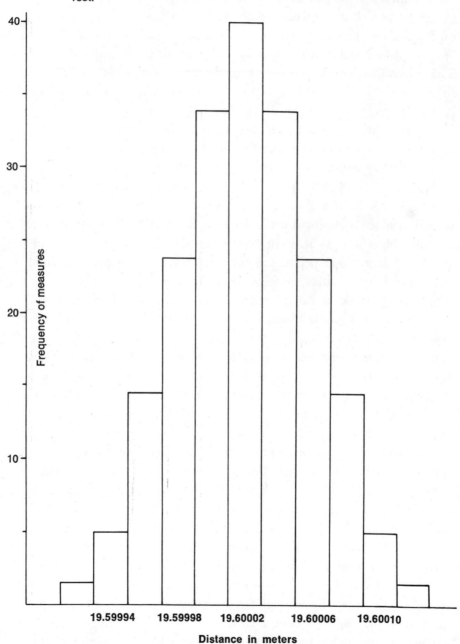

Figure 1—Estimated distribution of 200 measured distances traveled by a falling body in the first 2 seconds starting from rest.

dence available to me" leaves the scientist's statement open to revision based on new evidence. Similarly, interpretation of existing evidence may change. For example, you may be walking down the street and suddenly hear an explosive sound. You fall to the ground out of fear for your life. But as you lie there, on second thought you may decide it was a car backfiring. Note that the evidence has not changed, it has just been reinterpreted. Changes in interpretation due to reanalysis of the prior evidence, adoption of new interpretations from external sources, and new interpretations derived from added evidence are all quite common. Sciences, whether as well developed as physics or as recent and incomplete as sociology, have at least one characteristic in common: pronouncements are tentative. If they were not, we would have to deny the importance and even the possibility of new evidence and new thought. Such denial is contrary to the way of a scientist.

3. *Science is deficient because it cannot give any ultimate explanation of natural processes.* This charge is often made by persons whose understanding of the methods and objectives of science can most charitably be characterized as limited. The charge is true but irrelevant. Ultimate explanations simply are not part of the objectives of science in the modern world. A quest for ultimate explanations dominated much prescientific thinking. Thousands of years of effort and thought did not lead to a resolution of any ultimate problems, such as "What is the essence of life?" or "What is the greatest good?" In contrast, the preoccupation of science with potentially solvable problems such as "Why didn't milkmaids contract smallpox?" has produced spectacular results.

As stated earlier, scientific statements are tentative. An attempt at ultimate explanations based on tentative statements could be handled only by Don Quixote or Lewis Carroll's Red Queen. Beyond the logical problems, any ultimate explanation would automatically halt progress.

4. *Scientists distort reality and cannot do justice to the fullness of experience.* If distorting reality means that scientists select only one small portion of phenomena for investigation at any time, then they happily plead guilty. In fact, this limitation of the area of research is a vital tactic of science. Otherwise the problems are simply beyond our capacities.

In any given situation, there are literally innumerable things scientists could measure or manipulate. Achievement of any reasonable answers requires them to focus on one or, at best, a few aspects of any given situation. At the same time they attempt to minimize, in some manner, other factors that might influence the result.

The limitation of a discussion to certain aspects of the total situation occurs in any endeavor. If you are describing a baseball game you may discuss the size of the stadium or the color of the uniforms or the eccentric motions made by the pitcher or the number of hits each team got or the team spirit exhibited or where the fans are from or the temperature at game time or. . . . Or you may compare any of these things with previous games. But no one can discuss everything. So the scientist, like anyone else, selects the aspects of the situation he wishes to discuss. He differs from the nonscientist in being aware of the fact that he is selecting and in systematically investigating those aspects he has selected, although he may not be aware of how arbitrary his selection is.

To aid his investigation, the scientist in many instances attempts to minimize the differential influence of aspects he is not currently investigating. He does so by manipulating the environment and performing experiments. Criticism that science distorts reality often focuses on the laboratory experiment. Experiments are alleged to distort "reality" since the situation is artificial. Of course the object of the scientist is not to distort reality but to find out how the real world works by temporarily controlling events related to those he is currently interested in measuring.

The importance of such controls can be illustrated in the field of genetics. Genetics remained a matter of rather vague statements about "blood" and "like begets like" until Mendel and those who followed him examined the situation in carefully controlled and limited conditions. Mendel succeeded where earlier hybridizers failed largely because he studied inheritance of single contrasting characters, he carefully controlled other factors, and he recorded the results in detail. As you may have suspected by this time, science becomes sophisticated when it is able to create and observe artificial situations in which only certain factors can have an influence. The success of science depends on the ability to screen out the booming, buzzing confusion that confounds orderly observation.

In addition to being charged with distorting reality, scientists are also charged with failing to do justice to the fullness of experience. This accusation is most often applied to psychology, and it usually takes the general form that "People are too complicated and individually different to be studied through artificial laboratory situations." This argument is nonsense, of course. Scientists do in fact study man from the standpoints of both underlying biological processes and observable behavior. From these standpoints they create theories attempting to account for all aspects of human character. Their knowledge is both extensive and incomplete, as might be expected.

Each person is unique. So what? Every snowflake is reputedly different, yet scientists can still make meaningful statements about moisture, size, temperature, or crystalline structure. It is also said that every single atom is different. This statement may or may not be true. It is irrelevant! We do study atomic structure, but science has little or no interest in the completely unique. Its primary goals are to understand how observed facts relate to each other, to formulate models that approximate the relations of these observations, and to extend predictions to new observations. Few would object if you chose to spend a career studying a single snowflake; however, the scientific gain would not seem to justify this devotion. Not that the expression of uniqueness and aesthetic qualities is not worthwhile —it is just not a basic part of the scientific endeavor.

5. *Science is concerned primarily with man's practical and social needs.* The last of our misunderstandings is the assertion that science is man's reply to his practical and social needs. It would be fruitless to deny that practical and social needs play a part in science, particularly applied science. But they play only a small part; they are not the game. Unfortunately, these practical results of science are the most visible, so they are often identified as the heart of science. We know that the desire for these by-products is responsible for some scientific work and the financial backing of other work in the hopes of practical gain. For most scientists, however, science is a game played for understanding, not for practical solutions to existing problems. There is no really good evidence that simple need produces scientific advancement. If the scientific groundwork has been laid, necessity may encourage solution. For those scientists who play the game for understanding rather than practical advantage, it is a game whose chief delights are the addition of one neatly contrived stroke that helps give form to a picture; a game affording a glimpse of what no man has conceived before; a game from which may come ecstasy by bringing order out of chaos.

A simple glance at a few of the more important figures in science reveals the absence of "practical" goals. Galileo, Newton, and Einstein would certainly make any all-world science team. What practical need could have driven Galileo to astronomy, where he was criticized, rebuked, and threatened, both by academics and by the church hierarchy? Where was the practical pay-off? Could Newton or Einstein have been motivated by the hopes of future space travel or weapons that would fry dull uncomprehending masses as crisp as bacon? Utterly absurd. Nor does personal gain through either fame or money seem to have been their principal motivation. Both Newton and Einstein achieved fame at an early age but continued to work until their deaths. There is a story, which may be true, that Einstein, when invited to the Institute for Advanced Studies at Princeton, asked whether $3,000 per year would be too large a salary to expect. Of course there are those in science whose aim is a better mouse trap or personal gain. They may achieve visibility, but rarely do they make serious contributions to the basic concepts in science.

Science and the Application
of Scientific Information

The work of many individuals rests on a scientific base. Wide varieties of roles, training, and attitudes are represented. It seems useful to distinguish three groups: (1) those directly involved in what we speak of as basic science; (2) those who organize material based on the principles of science for application; and (3) those who apply the organized material to immediate problems. Very few individuals work exclusively either in basic or applied sciences; at times untangling the particular roles might not be worth the effort. In spite of overlap, however, there are sufficient distinguishing criteria to make this classification useful. So for present purposes we will assume that there are at least three reasonably distinct roles: the scientist, the applied scientist or engineer, and the practitioner. We can identify and classify these three science-related groups best by establishing a picture of their roles and relations. No single criterion would be adequate.

In our discussion of the three science-related areas, we will assume that all the problems attacked are currently soluble. There are many problems, of course, which at present cannot be conceptualized, tested, engineered, or treated for a variety of reasons. Resolution of a problem may be beyond our reach due to a lag in any one, or even in all three, of the groups considered.

1. The basic scientist is engaged in the task of articulating, deriving, and generating principles that hopefully have general explanatory power. His particular field of research may be rather narrow and his work may consist of attempting to fill in more specific detail within an already existing *conceptual schema.* That is, his work may depend on a set of interconnected ideas and propositions that provides a framework for consideration of theories. The conceptual schema contains both the explicit and implicit ideas that govern a scientist's thinking about a set of natural phenomena. For example, the conceptual schema may be those ideas associated with the assumption that higher organisms evolved from lower. "Mopping up"—investigating a part of nature in depth and detail within a conceptual schema —absorbs more scientific effort than any other single activity. Although this activity is not the sort of work that ordinarily results in a breakthrough, it does add clarity and precision to existing principles. For example, a great portion of the work of geologists over many years has been locating, plotting, identifying, and dating various formations. Techniques have improved vastly, but the greatest result has been a rather detailed determination of the structure of the earth's outer crust. The basic principles guiding this work were developed largely in the nineteenth century by William Smith and Sir Charles Lyell.

Scientists also engage in the sport of hypothesis testing. They predict from some set of assumptions and then test that prediction. For example, scientists have known for years that Cambrian formations (rock formations assumed to be five to six hundred million years old) contained rather highly developed fossil forms. From our knowledge of evolution it was predicted that Pre-Cambrian formations should contain simpler fossil forms. Unfortunately, for a long time Pre-Cambrian rocks did not yield any indication of prior life. But in 1947 fossil traces of simple form were found in Pre-Cambrian formations in Australia. Once the identification was made, other finds followed. The theory of evolution had once again survived examination by some of its most demanding critics.

The most exciting and confusing—but rarest—of scientific activities is the creation of new principles or conceptual schemas. These successful revolutions are not overnight affairs. The general pattern of revolution begins with mounting evidence that a current schema is not adequate. However, the inadequacy of a conceptual schema in and of itself does not necessarily produce a change. Rather, a revolution has to wait for someone to create a new conceptual schema. It is impossible to operate effectively without some framework. Better a rickety, worn-out conceptual schema than none at all.

To summarize, the work of a basic scientist is to organize natural processes through conceptual schemas and to collect data that test and give depth and detail to the schema. The basic scientist's primary task is to develop the concepts or principles that become more and more abstract as the science develops.

2. In contrast, the applied scientist is more closely identified with the *application* of principles and concepts to a specific and generally limited problem. The forms of application are so varied that no neat summary is possible. However, a couple of examples may illuminate the applied scientist's role. Among the most visible of current applications are those in the field of rocketry. To be sure, there are some new scientific problems involved and there has been some scientific "fallout" (that is, knowledge acquired that is incidental to the main objective), but the greatest problems have involved applications of previously existing principles and concepts. Consider, for instance, the velocity a satellite must achieve in order to go into orbit. What is involved? The gravitational field and the composition of the atmosphere are two major components of the problem. Each of these components has been the subject of prolonged scientific effort. In this situation the engineering effort that actually put a satellite into orbit used principles understood long before the attempt was contemplated. The development of the scientific principles concerning gravitation could, without too much distortion, be attributed to Newton. A more detailed knowledge of gravity resulting from "mopping up" gave engineers the approximations necessary. An appreciation of the composition and structure of the atmosphere goes back at least to Torricelli, a student of Galileo's. Torricelli's concept of a "sea of air" (about 1643) was critical to the development of orbiting satellites. This conceptual schema was developed in detail by succeeding generations of scientists.

The engineer attempting to design rocketry equipment faces problems of fantastic complexity. The difference between his situation and that of the basic scientist is that the problem is more specific (to put a satellite in orbit). The general concepts and details subsumed under these principles were previously available. The engineer asks "How does one use these principles to design a system that accomplishes the objective?"

A parallel example is the development of polio vaccine. Here again, the basic notion that many diseases are the product of specific viral or bacterial agents had long been known. In 1892, the Russian botanist D. J. Ivanovski laid the experimental foundation for research on viruses by transmitting the mosaic disease of tobacco using the cell-free juice of affected plants. By 1908, methods of transmitting poliomyelitis virus to Rhesus monkeys had been developed. Work on the problem of propagating poliomyelitis virus culminated with the critical discoveries of Enders and his associates in 1949. They managed to isolate and maintain cultures of the virus in tissue taken from human embryonic organs. These historical examples illustrate the sort of scientific background that often lies behind a well-publicized application. In addition to the basic knowledge about viruses, the principles of immunity were at least partially understood, and detailed knowledge of anatomy was available. The problem was how to use these principles to generate a solution for the particular problem. A very extensive amount of research in both basic and applied science was necessary before there was any possibility of developing a polio vaccine.

3. The third role is that of the practitioner. His work involves immediate contact with the materials, the patient, or the product. Thinking about general principles plays very little direct part in his day-to-day functions. His role is to use, in a single specific case, a solution generated by the applied scientist. For example, in the case of rocketry, a practitioner (here an electrician) might be involved in positioning the instrumentation that monitors the actual amount of thrust developed at a particular stage. Another practitioner may prepare the explosive bolts that separate burned-out sections of the rocket from the payload. A medical practitioner could be concerned with the recognition of symptoms, the proper administration of antibiotics, or other specific treatment using problem solutions previously generated by the applied scientist.

In addition to the somewhat specific differences noted above, the three roles can be differentiated along some general lines. The number of individuals involved in each of these roles varies greatly. The number of basic scientists is small compared to the number of applied scientists, while the number of practitioners vastly outnumbers both other groups combined. The level of abstraction required varies greatly from role to role, with the scientists' theories often far removed from ordinary sense data, whereas the practitioner deals with the here and now.

Generally, there is a time factor associated with these roles. The practitioner is faced with an immediate specific problem requiring an immediate specific solution. In most instances he can estimate quite accurately the time necessary to solve the problem. His problems often have known solutions; his primary task is to use them. If they have no known solution he compromises his results; he does not progress toward a general solution of the problem. The applied scientist ordinarily works over a longer time span on a single problem than does the practitioner. The applied scientist's specific problem has not been solved before, so time limits are harder to apply. He may or may not have a programmed deadline, but he is not generally required to devise an immediate solution. The basic scientist may be harried and driven by internal desires and interests, but there is literally no way he can promise solutions on a time schedule. In some instances he does not even know specifically where an interesting observation may lead, what the solution will look like, nor even whether he will arrive at one. Kepler took over 30 years to produce the three laws of planetary motion. Darwin's publication of *Origin of Species* came more than 20 years after his voyage on the H.M.S. Beagle.

A final difference in our three groups involves their source of livelihood. The practitioner ordinarily receives his salary or fees directly for his finished products, and he is in relatively direct contact with the consumer. The applied scientist's source of income varies considerably. He may be an independent consultant, he may be an employee of a firm that requires his talent in solving problems, or he may hold a faculty position in a university. With the exception of the faculty member, his income is directly allied to his problem area. The faculty member's time may be divided between the educational institution and outside employment, or he may work exclusively in a university setting, attacking problems simply because of their intrinsic interest.

Today the overwhelming majority of basic scientists are associated with universities. Financial support may be derived from faculty positions or from grants (ordinarily from the federal government) that support their research efforts. Most often their financial support has not been directly related to problem solutions for consumers. Disturbing changes in the relation of scientists to consumer problems have occurred in recent years, because there are growing pressures to work on immediate applied problems rather than basic ones. Evidence suggests that the greatest scientific advances are made when the scientist's theoretical problems dictate his research. Historically, the scientist has been the last to be supported financially for his particular work. In earlier times, some scientists (such as Benjamin Franklin, Francis Galton, and Charles Darwin) were wealthy amateurs. Others earned primary support from work not directly related to their scientific interests. Leonardo da Vinci's pathetic letter to Lodovico Sforza asking for employment emphasized his ability to contrive instruments of war as a basis for employment. Kepler served as court astrologer and often had to plead for payment of his back wages. Other scientists have been supported from sources widely divergent from their work. Francis Bacon was a lawyer and became Lord Chancellor of England. (This appointment terminated with his conviction for accepting a bribe. The result was that he spent the remainder of his life writing. For once, the law achieved a better reform than could be expected.) Lavoisier, the great French chemist, became farmer-general of taxes and, as such, paid the government a fixed fee, for which he was allowed to collect the state revenue. This profitable undertaking supported his scientific work but was cut short by the French Revolution; his scientific work ended at the guillotine.

In summary, we have seen some of the distinguishing characteristics of three science-associated groups. Overlap in some functions, attitudes, problems, and sources of employment is acknowledged, yet there are sufficient differences so that most individuals involved can be identified and do identify themselves with one of these groups. The primary difference among these three groups is their attitudes toward what is important in their work and the goals they hope to achieve.

Science and Pseudo-science

An introduction to science would be incomplete without some mention of Sunday-supplement or pseudo-science. These enterprises are often fascinating and amusing, but they are ordinarily different from acceptable scientific research. We must face the fact that the confused, misguided, or dishonest cannot always be distinguished from those who merely have new and, as yet, unaccepted views. The pseudo-scientist is well aware that many scientific innovations have been met with disbelief, scorn, hostility, and ridicule from more orthodox groups. Since the pseudo-scientist is often met with disbelief, scorn, hostility, and ridicule, he concludes he must be an innovator. (Again, the logic of the Red Queen.) Instances of rejection followed by overwhelming acceptance suggest that we should evaluate presentations carefully. These same instances provide some solace for the pseudo-scientist. How then do we improve our odds for identifying the pseudo-scientist while minimizing the risk of dismissing the merely eccentric or innovative?

The technique that has evolved to handle this problem is to allow both the pseudo-scientists and creative antiestablishment scientists to present their work. If the work has merit it may gain followers among those who happen upon it. Hopefully, a good creative idea would then snowball until it becomes a major scientific force while the pseudo-scientists' work would remain on the fringes.

Before we consider specific cases, let's recall two earlier statements. First, science is defined in terms of *how* and *why* we know something, not what we know. Second, scientific statements are judged by a jury of peers who are presumed to have substantial knowledge of data and experimental methodology.

We will consider three examples of varying degrees of nearness to science, ranging from obviously absurd beliefs to presumably well-intended but as yet unacceptable concepts.

By the beginning of the sixteenth century it had become less dangerous to believe in the earth as a sphere. Yet the belief in a flat earth persisted among some persons in the United States at least into the 1940s. A rather strange fundamentalist religious group held this view and defended it with what they considered evidence. According to "flat-earthers," the earth is flat and disc-shaped with the North Pole in the center and the South Pole ringing the outer edge. Sailing around the earth is merely describing a circle on this flat surface. The earth is motionless as well as flat, and the sun is 32 miles across and about 3,000 miles away. The fact that a person who jumps straight into the air for one second and does not come down 190 odd miles away was considered evidence that the earth is motionless. A picture taken across 12 miles of lake or a view of the opposite shore through binoculars was supposed to indicate that the surface is a plane. The "evidence" supplied by the flat-earthers was obviously highly selected to defend their position and did not take into account a myriad of facts that would refute it.

The second and slightly more sophisticated example of pseudo-science starts in dogma and ends in disaster. This one, mentioned earlier, is popularly known as Lysenkoism and involves the field of genetics. At an earlier time there were many substantial scientists, such as Lamarck and Darwin, who either accepted or toyed with the idea of inheritance of acquired characteristics. According to this theory, an animal such as an elephant originally had to reach up for food. This reaching stretched the trunk, and future generations inherited this elongated trunk. The rise of modern genetics and overwhelming experimental evidence deflated this idea by the early 1900s. Enter the Marxist dogma. According to Marxist fundamentalists, everything depends on the environment. The development of an animal or a plant is determined by its surroundings. Accepting this dogma, Lysenko concluded that plants grown under a particular set of conditions would acquire characteristics determined by those conditions, and the characteristics would then be transmitted to their descendants. That great scholar and scientist Joseph Stalin agreed and confirmed his position with brute force. Western observers visited Lysenko's farms, where he claimed to have achieved astounding results. These observers were overwhelmingly impressed by Lysenko's lack of understanding of controlled observation and his substitution of harangues for evidence. Showing patience beyond belief, they repeated Lysenko's experiments in other countries; the results were decisively negative. Despite this evidence, Lysenko held sway over virtually all Soviet genetics for about 20 years.

The third example perhaps brings us to the fringes of science. This is the supposed phenomenon of "extra-sensory perception" (ESP). The research to be considered is primarily that of Dr. J. B. Rhine. Dr. Rhine is certainly not in the same category as the "flat-earthers" or Lysenkoists. He has made a sustained effort to establish ESP as a scientific phenomenon, but after approximately 40 years of research and writing, his views and the ESP phenomenon have failed to be accepted by most scientists. His failure is primarily one of evidence and integration. The great majority of independent tests do not produce results supporting his concepts. In addition, his methods of analyzing data are simply not accepted by most other scientists. Probably his most fatal flaw is the inconsistency of his results. His subjects have a lamentable tendency to return to the statistical mean. Crapshooters are familiar with this sort of problem. Simply stated, it means that sometimes you will have "hot streaks" even with honest dice— for example, when 7 comes up five times in a row. But over a long period the shooter returns to the statistical mean, with 7 coming up an average of once every six throws. If one selects only the "hot streaks," it looks as though the shooter is controlling the dice. Selecting data appear to be basic to Dr. Rhine's interpretation. This is not selecting what kind of data to investigate, as other scientists do, but rather selecting "good" data from a pool that one collects. This second kind of selection clearly must be generally unacceptable to science. Even if Dr. Rhine and his colleagues are dealing with a real phenomenon, all they have been doing for 40 years is attempting to prove its existence. Scientists ordinarily integrate the phenomena they establish into a conceptual schema; Dr. Rhine's group has failed to do this. Another item creates difficulty for Dr. Rhine, too. This is the notion that whatever is being investigated is extrasensory; in other words, information is being transmitted by means that cannot be measured with any known instrument. Such an assumption is contrary to current scientific theories. This assumption, bizarre as it may seem, does not by itself rule out consideration of ESP as a part of science. The lack of acceptable evidence and interpretation of reliable phenomena are the fatal flaws.

Despite the obvious differences in the three examples, they have some common problems that leave them outside the accepted realm of science. The central problem is one of the quality of supporting evidence and the conceptual schema. In Chapter 2 we will discuss the nature of scientific evidence. For the moment, consider evidence in the same sense that a jury might. In a trial the burden of proof is on the party bringing the suit or making the charge; the charge holds a position similar to the new proposal. In science the burden of proof is, and rightly so, on the one challenging an accepted belief or attempting to establish a new conceptual schema. There is no requirement in science that we must attempt to disprove any particular assertion. The "flat-earthers" were completely unconvincing because their "evidence" failed on all counts. The assertions they made were highly selective, they violated the rules regarding what constitutes evidence, and, most important, their concepts simply did not take into consideration other existing evidence. Lysenko failed in that his "evidence" could not be supported by external sources; he failed in even a rudimentary understanding of how evidence is collected. He also failed to produce concepts that accounted for other well-substantiated observations. Lysenko was probably dishonest and was assuredly a political hack, although his character is irrelevant as far as the scientific worth of his work is concerned. Concepts and evidence must be judged independently of their authors. Finally, Dr.

Rhine's theory fails to be accepted primarily because of the inconsistency of the evidence.

Our major goal in this chapter has been to demonstrate that despite their agreement on the importance of science, scientists and laymen have very different ideas about what science is and how it operates. The critical difference is that the layman thinks of science as a collection of facts leading to practical ends, whereas the scientist thinks of science as a set of methods and conceptual schemas leading to an understanding of natural processes.

Chapter 3
Rules and Concepts

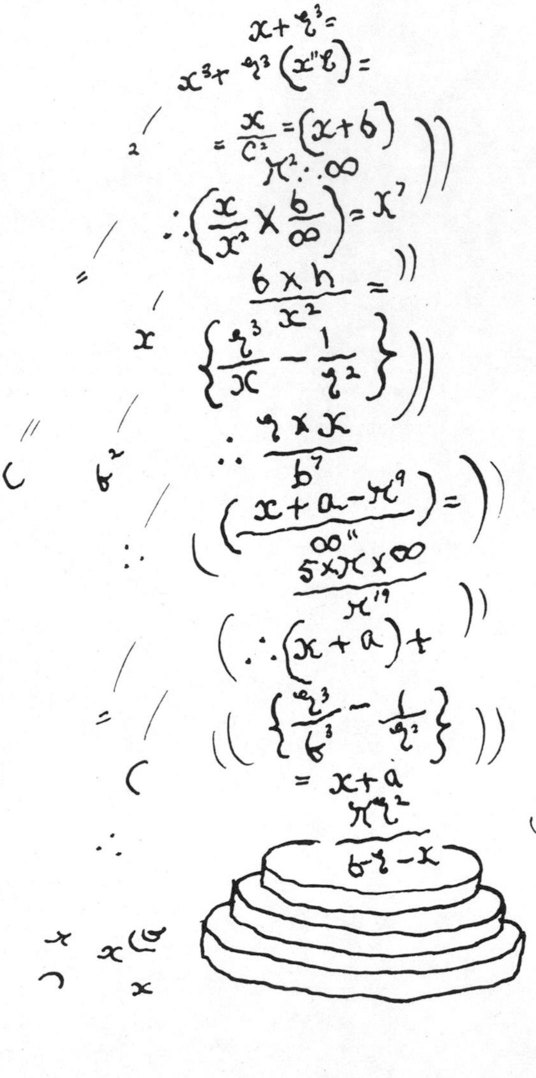

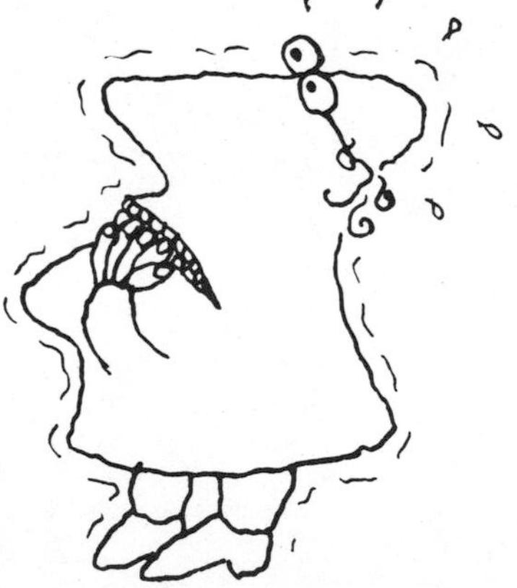

Something about scientific attitudes makes them seem significantly different from the other attitudes that have dominated cultures and subcultures. But what is it that makes science unique? This problem has been considered by scientists and philosophers for centuries, and there still are disagreements. Most of these disagreements are about the interpretations within science, not the methods of science. Even with these disagreements, scientists have substantially agreed on how to play the game. The person who plays has a clear idea of what is important in deciding whether a statement is true. The criteria for accepting or rejecting any statement as true are the logic of the arguments within a statement and the arguments' relationship to the data. The scientist may make mistakes or may be misled by false premises or emotions, as anyone may, but the rules for his decisions are well understood. The rules have evolved through centuries of effort by philosophers and scientists. They have thought about and tested many possible approaches to science.

A scientist, while playing the scientific game, makes no decisions on the basis of faith (he may make assumptions, but these remain open to question); no decisions on the basis of power; no decisions on the basis of monetary rewards; and no decisions on the basis of self-protection. A scientist has to be intellectually honest. The bases of his decisions as a scientist must be observed events and an attempt at their interpretation. It .is true that on some occasions scientists deny seemingly obvious and necessary conclusions about observed events and their interpretations because of preconceptions, but in doing so they are not acting as scientists. If someone makes a discovery that contradicts what he "knows" from his religious, economic, or other beliefs and summarily rejects this discovery, he is not playing the scientific game according to the rules.

Because of the importance of data in determining the beliefs of scientists, a tendency has developed to equate data and science. This misunderstanding was discussed in Chapter 2. A scientist's decisions and conclusions are validated by data, but the data per se are not the science. The scientist gathers data and studies them, but in order to make the whole enterprise scientific he must also organize and intrepret the data. A more specific discussion of how data are organized and interpreted appears later in the book.

Two people can hold the same beliefs and have the same knowledge. One may be a scientist and the other not. For example, if all of the scientific information known today were to be studied and accepted with no new work done, there would be no science. If in answer to the question "How does energy relate to matter?" someone reads that $E = mc^2$ and learns that "energy in joules equals the mass converted in kilograms times the speed of light in meters per second squared," he is not necessarily playing the scientific game. A statement might look like science and sound like science, but it is not science if the only reason one believes the statement is that he has read it or Einstein has said it. Having knowledge is not what makes a scientist. It is the *method* of attainment of knowledge that determines whether one is playing the game according to the rules.

Aristotle was a brilliant man; he is remembered primarily as a philosopher and logician, but he was also a scientist. He observed events and thought about them. He made many statements related to these observations. The truth or falsity of these statements is not critical in determining their scientific status; only their method of attainment is critical. The acceptance of these statements during the late middle ages by the Scholastics was not scientific, because for them truth depended solely on the fact that Aristotle made the statements. They did not justify their belief by reference to data.

Science versus Dogma

For someone playing the game of science, the method by which he
ascertains what he believes is crucial. He has to evaluate data and arguments and decide for himself on their validity. A scientist communicating
to others has the task of convincing the hearer of the validity of his
statements in terms of the data and their explanation. He is not playing the
game correctly if he wins support by the strength of his personality or
prestige. It is up to the hearer who respects the scientist to evaluate what
the scientist says, rather than accept it because he says it. And the culmina-
tion of this enterprise is the determination of whether the explanations
account for the data observed. The relationship between scientific explanations and data helps to ensure the integrity of the scientist. If his
findings are important he can be sure that his theories and data will be
examined critically and his experiments repeated.

It is the system of data-based explanation that distinguishes science from
dogma. The scientist has both the right and the responsibility to decide for
himself, on the basis of the evidence at hand, the best explanation of a set
of phenomena. He also has the right and the responsibility as a scientist to
investigate thoroughly the bases of his beliefs. He cannot accept statements
unsupported by data. On the other hand, dogma (religious, economic,
political, social, or any other kind) depends on pronouncements by established authorities (for example, the dogma that the earth was created in
4004 B.C.). The goal of the student learning a dogma is to accept the pronouncements as they are given to him. If he disagrees with the dogma, he is
not playing the game of dogmatism correctly. It is his right and responsibility to believe the dogma. He has to search his soul until he accepts it or be
considered an outcast and suffer the consequences. It would not matter if
he could present strong arguments in support of his personal beliefs. In
dogma, arguments and facts are forced to coincide with the dogma. The
student cannot accept statements that do not agree with the dogma.
(Continuing the previous example, he must reject the existence of prehis-
toric man around 10,000 B.C.)

One way of contrasting science and dogma is to say that a scientist
accepts facts as given and belief systems as tentative, whereas a dogmatist
accepts the belief systems as given; facts are irrelevant.

In fact, it is possible that dogma exists in the realm that is normally

thought to be scientific. For example, if a physicist accepts the theory
that a person who was accelerated away from and back to his twin would
be younger than his twin when he returned (the twin paradox) because
Einstein said so (rather than by understanding the arguments and data—
if any—that confirm it), he is acting dogmatically. In fact, in *Physics
Today*, Sachs argues that there are no data supporting it and that the
arguments of the twin paradox are fallacious. Reader response treated him
like a heretic rather than a physicist with another opinion. It is difficult
not to be dogmatic about one's beliefs, whether they are justified or not.
The theoretical statements of many scientists have been made dogma by
their followers. That is, their followers accepted these scientists' statements as absolute truths. Such unfortunate scientists include Aristotle,
Ptolemy, Lamarck, and Freud.

Basic Concepts Involved in Playing the Game

We now turn to the details of playing the game of science. What counts as
data and what counts as an explanation of data? These are questions that
have controversial answers. Scientists do similar things but they don't
always evaluate what they do in the same way. One major reason for this
disagreement is that one cannot in any satisfactory way separate data from
explanations of those data. Attempts have been made to separate them, but
no formal analysis of the difference between data and theory has been
satisfactory to all concerned.

Your understanding of the game of science may be aided by a brief
description of facts, data, laws, and explanations, along with a discussion of
their similarities and differences.

Scientific Facts and Scientific Data

There are disagreements among scientists over what constitutes a scientific
fact. The term is regularly used in at least two ways. One way is to call a
well-established law a fact. (Laws are discussed in the next section.)
Another is to use the term "fact" to refer to the actual occurrence of an
event. The latter definition is the one used in this book.

The term "fact" and the term "datum" get confused when a specific
event is referred to. The *fact* is the actual occurrence of the event; the

datum is the recording of that event. In other words, the datum is the representation of the fact by some means that is relatively permanent, whereas the fact is an event that occurs and then is gone forever. The datum is usually recorded in a symbolic form (words or numbers) at the time the event is observed.

Consider the following four statements:

1. There are 12 books on Professor Brown's desk right now.
2. There was a thunderstorm in Chicago last night.
3. The temperature outside is currently 83° Fahrenheit.
4. He traveled 262 miles in his car between the last two times he filled the gas tank, and on the last occasion he added 11.8 gallons of gasoline.

As written, these statements are data. The events they refer to are facts. As data, they are very incomplete because they do not accurately identify the specific events they refer to.

According to the definitions we are using, the following list does not represent facts, but rather laws, generalizations, or inferences.

1. Professors keep many books on their desks.
2. It rains a lot in the springtime.
3. It is warm in the afternoon.
4. His car gets approximately 22.2 miles to the gallon.

The major difference between the two sets of sentences is that the first set describes *individual* events and the second set describes *classes* of events. According to this definition, facts are singular events that occur at a given time. However, in order for an event to be admissible as a scientific fact it has to meet other basic criteria. The first criterion is hard to formalize but is very important. The event has to be one of a type such that, in principle, more than one person could describe it. This criterion is known technically as *intersubjective testability*. The current position of most scientists is that nothing is admissible as a scientific fact unless it is intersubjectively testable. One way of assuring intersubjective testability is to require that scientific facts be describable ultimately in terms of tables, chairs, colors, sounds, pointer readings, weights, and so on. That is, they have to be describable in the language of physical things (called the "physical thing language" by some philosophers). Sensations and pains are not so describable, although the *report* of sensations and pains is. This criterion of intersubjective testability is an important one, and one that is generally accepted as necessary. It means that what you see in a dream is not admissible as a scientific fact, because no one else can see your dream. If you have an itch or a pain, that is not admissible directly as a fact. If you take LSD or mescaline and report wild or horrendous sensations, the wild or horrendous senations are not, in and of themselves, admissible as facts.

Psychologists will accept as a fact your report of a dream, a pain, or a wild sensation; but a private sensation, although very real, is not itself admissible as scientific fact. If someone says "I dreamed that the moon fell toward me while I was lying in bed," the utterance may be considered a scientific fact. The recording of the utterance may be considered data. The experience of the dream is neither, because it is not, in principle, observable by others. The dream itself can enter scientific discourse only as a theoretical concept. It is not itself observable, but if hypothesized can lead to observable consequences, such as a change of EEG patterns, rapid eye movements, as well as verbal reports of bizarre experiences upon awakening.

A second criterion for an event to be admissible as a scientific fact is the requirement of a high degree of agreement among different people on the description of the event. This criterion of *reliability* is related to but somewhat different from the criterion of intersubjective testability. Intersubjective testability requires the event to be open to public view and not entirely personal. Reliability requires the event to be described in such a way that different individuals can agree on the description. It is this description of events, not the events themselves, that scientists enter as datum, although to simplify communication many scientists refer to the datum as the event rather than its symbolic representation.

Most events can be described so that different individuals will agree on the description. For example, one might simply say that an event occurred. Since everyone would agree that an event occurred, that description is highly reliable, although it is not very informative. Reliability has associated with it a criterion of *precision*. The more accurately and specifically an event can be reliably described, the more acceptable the description is to scientists. Not only should the event be described so that different individuals can agree on the description, but it should be described in such a manner that the event can be differentiated as much as possible from similar events along relevant dimensions. You may describe the temperature of a bowl of water as warm or cool, and other people may describe it in a similar way. Another person's verbal description "warm" describes a range of temperatures that probably is relatively similar to but different from yours. You can make the description both more reliable and more precise by using a thermometer and describing the temperature as 77° F. By describing the temperature in this manner, you differentiate the event much more precisely from other events, and different individuals can more readily agree on the accuracy of the description. Precision is one of the goals of scientific description of events.

To help improve reliability and precision as well as for other reasons, some of which are obsolete, certain philosophers and scientists introduced the concept of operational definition. Like the other criteria, this one cannot be formally defined, but it can be discussed and used. An operational definition is the description of how one gets his fact. For example, an operational definition of the temperature might be "A mercury thermometer that was in the shade, over grass, four feet off the ground, in the open on three sides and above, gave a reading of 93° F at 2:00 P.M." The description cannot be totally complete without being extended forever, but many relevant variables are mentioned. When a scientist writes, he implies many things without stating them, and he ignores other things because he thinks they don't make any difference.

Like the mother who didn't say to her children "Please don't eat the daisies," our operational definition did not say "There was no fire burning nearby." It also didn't say "The thermometer was 10 inches long" or "The thermometer contained 1.7 grams of mercury." As you can readily see, possible descriptions of the conditions under which data are collected are inexhaustible, so no operational definition is complete. Despite its incompleteness, however, an operational definition can improve specificity in scientific communication.

In summary, there are three main criteria applied to events for judging whether they are admissible as scientific data: (1) The event that is accepted as a scientific fact must be singular. Data are symbolic representations of singular events. Interpretations and generalizations given to these events are not themselves data. (2) The event must, in principle, be available for public scrutiny. It does not necessarily need to be sensed by more than one person, but it must be of a type that could be sensed by more than one person. (3) The description of the event should be such that different individuals can know, as specifically as is reasonable, what the event was that is being described.

Scientific Laws

Although scientists use individual events for data, their explanations always rest with classes of events. There are many classes of events, however, within which observations are so well established that a *law* or *generalization* is accepted by almost all scientists. For example:

1. The speed of light is a constant of approximately 186,000 miles a second.
2. All cultures have a form of the family as a principle institution.
3. Oxygen is about 16 times as heavy as hydrogen.
4. The gestation period of elephants is about 2 years.

These laws or classes of events are what many people erroneously call science. They are the "facts" that people think of as the defining characteristic of science, although, as we have seen, just knowing these laws is not the most important thing in playing the game of science. Laws do not define science, but they are some of the most important products of science. They are the principles used by applied scientists and practitioners in solving specific problems.

What, then, are scientific laws? Laws are statements describing different properties that go together in the same kind of event or in certain sequences of kinds of events. They are descriptions of relatively constant relationships between certain kinds of phenomena. These descriptions may be in sentence form or they may be in symbolic form, such as an equation. It is important to remember that laws do not refer to singular events; rather, they refer to *any* singular event that has the properties defined in the law. This last characteristic of laws is the major reason that Dr. Rhine's research doesn't qualify. He has no prior criterion of which data follow the "law" of ESP and which do not.

Scientists expect all events having the necessary properties to conform to the law. As indicated before, specific measures of events lead to probabilistic statements. On this basis we would not expect any single measurement to conform to the law exactly. But scientists do expect relationships between kinds of events to be relatively consistent. It is the reproducibility of phenomena that leads one to accept a relationship as a law. The fact that any object starting from rest in free fall in a vacuum shows the approximate relationship $S = \frac{1}{2}gt^2$ leads one to call this statement a law. The law that oxygen is about 16 times as heavy as hydrogen was accepted, among other reasons, because decomposition of water regularly yields about two volumes of hydrogen to one volume of oxygen and because the oxygen produced always weighs about eight times as much as the hydrogen produced. Laws are established by the consistent repetition of relations between kinds of events; not by a singular occurrence of any succession of events.

Problems arise for scientists when an event that is supposed to obey a law does not. We cannot punish the event; like the customer, the event is always right. If an event does not obey a law as it should, it is the law that is at fault, not the event. It may be, of course, that the scientist made an error in his experimental setup and the event did not have the characteristics stated in the law. There may have been errors of measurement or of interpretation as well. If a law has been repeatedly confirmed, a scientist is usually very reluctant to modify it, so he is likely to check carefully to see that no errors were made before he attempts such a modification.

So far, this discussion of scientific law is quite abstract; let's look at a few short examples. It is a primitive scientific law that things fall. This law means that on earth any event having the property of thingness moves in the direction of the center of the earth when support is taken away. Like all others, this law has specific applications and limitations. A child's balloon when filled with helium does not fall; it rises. So a qualifier must be made. After much experimental research, scientists concluded that "support" includes air. Air is a fluid and has weight; therefore it can support anything that for a given volume weighs less than air. Just as a boat sinks until it displaces water that weighs as much as it does and no more, a balloon will rise until it displaces air that weighs no more than the balloon. If the supporting air is removed, the balloon will fall. Note that a scientific law depends on many conditions, not just a single one. Theoretically, a scientific law holds for all events that have the right set of properties.

Continuing the same example, one might ask, "What about birds and airplanes? They don't fall all the time." The answer to this query is a little more complex, but essentially it again leads to a limiting of the conditions under which the law holds. Certain surfaces, like wings, when moving through the air set up a partial vacuum which, in effect, increases the volume of air displaced until the weight of the air displaced is comparable to the weight of the winged object. Thus birds and airplanes would fall except for the fact that in flight their wings generate a force to counter-act the fall.

Another example of a law is that living organisms are composed of cells. In other words, anything that can be defined as a living organism has as its building blocks one or more relatively self-contained units known as cells. This law states that if an event has one set of properties (for example, "living") it is expected to have another set ("cellular structure"). This particular law was established as universal by a biologist, Theodor Schwann, in 1839. Like falling bodies, this law requires elaboration and identification of limiting conditions. Certain aspects of an organism may not be cellular because cells secrete noncellular material (hair, hormones) which remains in or attached to organisms. Also, a problem with this law may exist with viruses that are not cellular (they are protein complexes), but then they may not be living organisms.

To summarize, there are four criteria applied to any statement before it can be accepted as a scientific law: (1) The statement must be about kinds of events and not directly about any singular event. (2) The statement must show a functional relation between two or more kinds of events ("kind of event" refers either to things or to properties of things). (3) There must be a large amount of data confirming the law, and little or none disconfirming it. (4) The relation should be applicable to very different events (although there may be limiting conditions).

Scientific Explanations

We have seen examples of facts and of laws or principles. We shall now consider some examples of explanations. Developing and elaborating explanations is a major part of the scientific enterprise. The problems of defining facts and laws are difficult (and we did not resolve them), but the problem of explanation is even more so.

A scientist tries to take the data in a given area and invent a general principle or set of principles with which these data are compatible. In other words, he attempts to develop a framework within which he can view events and data and understand them. In general, the greater the number and diversity of events that can be explained from a small set of principles, the better the explanation. An important part of the game of science is to develop the smallest set of hypotheses or principles that will account for the greatest variety of events. Once the scientist arrives at seemingly workable principles, he then reverses the process and uses these principles to indicate new facts to be observed so that he can find out whether these new facts are consistent with the proposed set of principles.

Consider the work in astronomy. People have observed the heavenly bodies since antiquity. Anyone who casually observes the skies and notes what he observes will find out certain facts and laws. On one night the observer will notice that the stars move gradually toward the west as the night progresses. On the next night he will notice again that the stars move gradually toward the west. If he continually watches the heavens he becomes aware that the same patterns of stars appear regularly, and furthermore he notes that they do not appear at the same place at the same time. Each night a given pattern is at the same place in the sky about four minutes earlier than the previous night. Almost everyone who looks at the skies assumes that the same pattern means that the same stars are appearing again and again. Our observer notices peculiar stars, however, that appear near the same place among other stars on successive days, but in the course of a longer period of time change their positions quite radically. These stars have been called "wanderers" or "planets." All of these observations and interpretations were made by the Egyptians and Babylonians, and they serve as a background set of beliefs and information necessary to our understanding of later developments.

As an example of an early attempt at explanation, consider Aristotle's discussion of these events. He suggested that the earth was a stationary sphere in the center of the universe. All the heavenly bodies were on huge, clear spheres that went around the earth in perfect circles. The sun, moon, and each planet had its own sphere spinning around the earth, and all the fixed stars were on a single sphere farther away from earth than the others. The different motions of the planets were explained by the assumption that different spheres all traveled around the earth at different velocities. All the heavenly bodies were perfect and unchanging, and each of them traveled along one or more perfect circular paths.

This is an explanation of the facts as they were known to Aristotle. From it, one can ask certain questions and get answers. (1) Why do the same patterns of relations generally exist among the stars? Because most of the stars are spots on the same sphere. (2) Why is it that five stars, the sun, and the moon do not remain in the same place among the other stars? Because each of these seven heavenly bodies has its own circular orbits revolving around the earth at their own rates, which together are slightly slower than the stars' revolution. The sun makes one complete revolution around the earth a day. The stars also make approximately one revolution a day. However, since on a given night they are about four minutes more westerly in the sky than on the previous night, they make one more revolution a year than does the sun. The moon makes about 13 revolutions fewer than the sun; and each planet goes around on its own schedule. From this theory, by knowing where a heavenly body is at any given time an observer can predict where it will be at another time and he can postdict where it was at a previous time.

Aristotle's theory was not very powerful because each orbit had to be independently plotted and there was no general explanation of why the planets orbit at their particular rate. Furthermore, there was no indication of the nature of the huge sphere that housed the stars and was the ceiling of the sky.

Using this theory or similar ones as a basis, however, astronomers tried to plot more exactly the paths of the planets to establish how fast they traveled in their circular routes around the earth. These astronomers found problems developing out of these explanations. Sometimes planets seemed to be traveling faster than the fixed stars, although generally they traveled more slowly. Two of the planets, Venus and Mercury, never got very far from the sun. When they rose in the east before dawn they were known as morning stars; when they remained in the western sky after the sun went down they were known as evening stars. From these observations of the planets, the astronomers concluded that none of them could be circling the earth at a constant speed. Simple circular explanations were not consistent with the observed facts. When an accepted explanation no longer explains the major facts well, scientists recheck them and some usually try to modify the theory to handle the facts. Aristotle modified the theory by adding rotations within rotations. In this way he could still explain the heavens in terms of constantly moving spheres. However, another possible modification could have been to assume that the planets did not travel at a constant speed. Sometimes they would move faster than the stars and sun, sometimes at the same speed, and sometimes slower. However, simply stating that the planets change speed is not a satisfactory explanation, since one needs an independent reason why the planets would speed up or slow down. When an explanation becomes solely a restatement of the data, it is not a scientific explanation. We are simply giving the phenomenon a name. Our example leads only to questions and answers such as "Why is Venus rising earlier and earlier daily? Because it is moving faster." "Why is Venus now rising later and later? Because it is now moving slower." These answers can be reduced to, "It is moving faster because it is moving faster and it is moving slower because it is moving slower," which is no explanation at all. An explanation has to be a conceptual schema that organizes and extends the data; its truth or falsity has to be testable. In other words, if all *possible* facts automatically confirm a theory, the theory is worthless. Some possible facts should tend to negate a theory.

Consider a minor theory (the term "theory" is here being stretched to make a point): "The television set in my house is on." You can confirm this theory by observing the set, seeing a picture, and hearing the sound. You can negate it by looking at the set, seeing a blank screen, and hearing no sound. Since the theory can be negated it is meaningful.

Consider another minor theory: "The television set in my house is either on or off." You can confirm this theory by observing the set and seeing that it is on. You can also confirm the theory by observing the set and seeing that it is off. There are no other possible observations. Therefore the theory is worthless.

A later explanation of astronomical phenomena—one that was accepted without question from the second century until modern times—was that of Ptolemy. He modified Aristotle's position by claiming that the planets traveled in little circles (epicycles) on their orbits while they were circling the earth. Explaining the paths of the planets this way enabled Ptolemy to predict pretty well where the planets were likely to be at a given time. For accurate predictions, Ptolemy had to assume that the center of the orbit of the planets was not always the earth, but he arrived at no reasonable explanations for the eccentricity of the orbits.

During the sixteenth century, Nicholaus Copernicus had a major insight. He studied and plotted the paths of the planets, but he could not fit the data (the planets' locations over time) to the Ptolemaic system. He attempted to explain the data by assuming that the sun, and not the earth, was the center of the universe and that the earth went around the sun and spun on its axis, as did all the other "wanderers" except the moon. By assuming that the earth spun on its axis once a day, he could hypothesize that the fixed stars that seem to rotate were actually stationary. They would no longer have to be on a huge curtain that spun at a fantastic rate. Day and night could be easily accounted for by the earth's spinning on its axis. In fact, this theory had unexpected explanatory power; the year and the seasons could also be accounted for easily, and the planets could again be plotted as traveling in simple patterns: their seemingly complex path was due simply to the position of the earth when the planet was viewed.

Copernicus' explanation, which came after much hard work, directly contradicted both the religious dogma and the scientific explanations of the day. This explanation revolutionized man's thought and was vital to the new learning of the Renaissance. The explanation, however, was not completely satisfactory.

About 50 years after Copernicus died Johann Kepler began trying to find a simple way to describe the paths of the planets. He spent years studying the data of Tycho Brahe and others and personally plotting the path of Mars. He assumed that the sun was responsible for the planets' paths, and that they revolved around it. But a circular path did not fit the data. Mars was not always the same distance from the sun, and putting the center of the circle elsewhere made no sense to Kepler. He tried many different ovals with the sun at the focus and still he could not fit the path of Mars. Sometimes the planet's orbit was inside and sometimes outside of the ovals he tried. He finally tried to approximate the orbit with an ellipse. To his surprise, he could now predict the path without error (within the accuracy of his measures) assuming the sun at one focus. This prediction, after much more contemplation, led to Kepler's first law of planetary motion: planets travel in elliptical orbits with the sun at one focus. The data collected about other planets were consistent with this law. Given reliable paths for the planets, Kepler could now formulate a law to account for their relative speeds in their orbits. This law is: A line connecting the sun with a planet sweeps equal areas in equal times. The planet does not travel at a constant speed but travels faster when nearer the sun than when farther away. His third law—that the ratio of the cubes of the mean distances of any two planets from the sun equals the ratio of the square of their periodic (orbital) times—delighted Kepler and showed the power of the explanation because it accounted for more than did any previous theory. It gave a systematic relationship to the speeds of the different planets, and it explained why some planets traveled around the sun faster than others. From these three principles, with only a limited amount of data, one could generate the paths of all the planets and consequently explain the observed movement.

In addition to being descriptive *laws*, Kepler's laws are *explanations* of the motions of a planet because he relates the motion of a planet both to the motion of other planets and to its specific relation to the sun. In answer to the question "Why does Mars travel in an elliptical orbit?" we can answer that all planets travel in elliptical orbits. If we ask "Why does it travel faster sometimes and slower at other times?" we can answer that the planet increases its speed as it nears the sun, since the radius vector (a line extending from the sun to the planet) sweeps equal areas in equal times. To the question "Why does Mars travel faster than Jupiter?" we can answer that the closer the planet's orbit is to the sun the faster the planet travels, and we can even formalize the specific function relating the two planets.

Explanations are never complete; Kepler did not explain why planets traveled in elliptical orbits, nor did he explain why their radius vector swept equal areas in equal times. But Kepler did offer legitimate explanations. The law "Mars travels in an elliptical orbit around the sun" is part of an explanation of why Mars is seen where it is in the sky on a given day. The law "Planets travel in elliptical orbits" is part of an explanation of why "Mars travels in an elliptical orbit." The law "The closer the planet to the sun the faster it travels" is part of an explanation of why Mars travels faster than Jupiter and also why Mars travels in its orbit at variable speeds. Kepler thought that the sun caused these behaviors, since it is the focus of the radius vectors. It can be seen that laws are used to explain lower-order laws, and these laws themselves then become candidates for explanation. Kepler's laws were explained by Sir Isaac Newton.

Newton, who was born 13 years after Kepler died, explained Kepler's laws by his gravitational principles. Whereas Kepler had three laws of planetary motion that explained the movement of all the planets, including earth, Newton had three laws of motion that related such extraneous things as falling apples and the tides to the same principle as planetary motion. From Newton's principles, not only could Kepler's laws be derived, but deviations from the elliptical orbits (too small for Kepler to see with his instruments) were also accounted for.

We have seen that scientific explanation, which is at the core of the game of science, requires relating the laws describing a class of events to some set of principles. When the laws are seen to be derived from the principles, then scientific explanation is achieved. It must be noted that scientific explanation is more than a description of events, and it is more than a law describing the particular class of events. Explanation requires an interpretation of what the events are and an indication of the reason they behave the way they do, by relating them to a different set of events. The scientist investigates carefully to see whether the events behave the way the explanation suggests they should, and he investigates other events that should behave the same way. Generalization to a different set of events cannot be done if the scientist only describes and does not give an explanation. In astronomy, a description is about spots of light that appear in the sky nightly, whereas an explanation is about solid bodies traveling around another body in elliptical orbits due to the gravitational field surrounding all bodies.

Obviously, as a science advances, new explanations develop in an attempt to explain the same kinds of events. But there is an important consideration in this advance. Although astronomical events were essentially the same in Aristotle's time as in Newton's, the data were not. As the technology of observation of the stars improved and more people systematically plotted their movement, the data changed considerably. A good scientific explanation is one that is in accord with a great amount of data. The data abstracted from the events are the test of an explanation; the events themselves are not such a test.

We can see from the preceding lengthy example that men have been fascinated by the heavenly bodies for over 5,000 years. Why this interest? They had a puzzle. The puzzle was laid out above them every night. It naturally piqued their curiosity and challenged their ability to solve it. Were there any other reasons? Probably. Does it matter?

The Public Nature of Science

We have seen that science depends on an individual's observing facts, collecting data, and coming to his own conclusions about how to organize the data. However, it is also important that the individual scientist communicate his findings and explanations to others. Although a single individual living on a desert island could do science, science is not a game of solitaire; it has many participants. A scientist's explanation does not aid the science until he can convince other scientists that he has an explanation. Not only must an individual make up his own mind; he must also convince other scientists that his explanation is valid.

The reasons for the necessity of convincing others of a position are threefold. First, by attempting to present data and argue for an interpretation, the scientist has to spell out in a clear manner what the data are, what the explanation is, and what the reasons are for believing the explanation. The scientist himself may gain new insights from this intellectual activity. In spelling out a position, a scientist on many occasions finds new and unexpected problems in his explanation, and he may thus find further work necessary to clarify his position. In order to present a position to others, a scientist must evaluate his position objectively. Insights accompanied by feelings of exhilaration are not enough. They are not always associated with valid explanations. During an objective evaluation and formulation, the scientist may convince himself that an explanation which he thought valid was not. Of course, evaluation does not always lead to disenchantment with the tentative explanation. The scientist may, to his delight, find that he can extend the explanation into unexpected areas, or that slight modification will enable him to do so.

Second, the output of a scientist must be put on public display before it can be considered a scientific contribution. This requirement makes scientists careful in their formal theorizing, since an ill-conceived or poorly supported hypothesis can be (and often has been) a lifelong embarrassment when it endures in cold, unerasable print. By the nature of the game, a scientist's work has to be open to public scrutiny; the only method of gaining acceptance requires having one's work approved by his scientific peers. If an individual presents his data and explanations but cannot convince some peers of their validity, the scientific community as a whole will tend to reject his claims. The scientific community requires an explanation of data to be acceptable to a sizable minority of that community before the privileges of scientific consideration are extended, such as publication in scientific journals and inclusion in textbooks and monographs. Without this exposure, a theoretical position has little chance of directly contributing to the science itself. Although such restrictiveness of science is sometimes in error, allowing important contributions to be dormant for many years, it does ensure that journals and textbooks contain significant material.

The third major reason for communicating scientific discoveries is simply to inform others about them. Convincing oneself and others of the validity of the discovery may be important in the advancement of one's own knowledge, but society at large should also be informed. Only after scientific knowledge is communicated to applied scientists and practitioners can it be used to benefit society. If the principles of electromagnetic waves had not been known to Marconi, he would have had great difficulty in developing the radio. If the principle of universal gravitation had not been known, it could not have been used to develop artificial satellites. Also, without free and open communication different people would spend years working on the same problem. There is a great waste of both effort and scientific ability if a scientist has to go through the same steps that someone else has already been through. Finally, only through free communication in books and the mass media can the general public learn about themselves and the world. In the modern world, the general public indirectly supports scientific research. This public should know something about what it is buying.

Scientific Methodology

We have seen that science has as its object the natural world, and that scientists strive to explain the facts of nature. We have discussed briefly what constitutes a fact and what constitutes an explanation. Our next problem is to examine how the scientist goes about finding facts and establishing explanations of them.

When a scientist is at work he is not likely to gather facts by casually observing the world around him. He may start by casual observation, but facts obtained in this manner rarely show the relationships necessary for a powerful explanation. Such an explanation requires precise and reliable observations. Casual observation is likely to be contaminated by many extraneous factors.

Obviously, the explanation of planetary motion established by Kepler could not have been done without carefully controlled observation. The information used by Kepler had to be detailed and accurate. In addition to all of his hard work, Kepler had very good luck. In planetary motion, the effects of all variables except those caused by the sun are quite small. The elliptical orbit, which describes the path that a planet would take if the sun were the only influence on planetary motion, is the dominant orbit. The gravitational fields of the planet itself and those of other planets are quite small compared to that of the sun. By using the observations of Tycho Brahe and observing as carefully and objectively as he could, Kepler plotted the orbits of the planets, and with hard work and insight formulated his major laws of planetary motion. The naturally occurring orbital paths are almost identical to Kepler's theoretical ones.

Aristotle, another great scientist, established many of his scientific laws on the basis of casual observation. One of his great laws was that heavy things fall faster than light ones. This generalization, which had much observation to support it, was accepted as a fact for almost 2,000 years. Anyone knows that rocks are heavier than feathers and that they fall faster. In fact, feathers don't even fall, they float. One problem with Aristotle's law is that falling bodies are affected by many things. In a very famous demonstration (perhaps apocryphal), Galileo went to the top of the Leaning Tower of Pisa, and before a large crowd simultaneously let a heavy and a light ball fall. To everyone's (except Galileo's) surprise, the balls stayed next to each other during the fall and landed together. By setting up an artificial situation, Galileo provided data that allowed him to establish an important natural law.

A paradox within science is that artificial, or controlled, situations help scientists understand the natural world. In natural settings many aspects of any two events differ from each other; the scientist sets up artificial situations to make different events similar to one another in many ways. Consider the example of falling bodies. Many events influence their fall— the impetus with which they start falling, the things that divert their fall, their shape, the amount of air they displace, as well as (possibly) their weight. Also, the difference in their rates of fall may be slight. If two things fall at different times from different heights, an observer may not know which actually fell faster. Even if he does, he may not know whether the height from which they started to fall made any difference. Galileo contrived the time, place, and condition of the fall; only by doing so could he discover certain natural laws. By dropping two balls of different weight from a great height at the same time, he controlled the height of the fall, the time of the fall, the shape of the objects, and the material of which they were made. By letting the objects fall a long distance at the same time, he could determine even slight differences of speed. When they fell together, he could easily conclude that weight per se did not influence the speed of falling objects. Observations in natural, or uncontrolled, settings never led to this conclusion.

To study this phenomenon further Galileo built long, smooth, inclined planes and rolled balls down them. He could have watched logs and rocks roll down hills, but again many factors would have been different at different times. Galileo worked for years carefully releasing balls and timing how long it took them to cover different distances. By controlling the situations—that is, making them artificial—he could determine and study the important relations. Like Kepler, he followed false leads and had trouble establishing simple relationships, but he finally came to important conclusions concerning motion. One conclusion was that a constant force does not result in a constant velocity; rather, it results in a constant acceleration. Time and acceleration were seen to be more important variables than velocity and distance. These important discoveries could not have been made if Galileo had not invented artificial situations in order to control and accurately measure the effects of important variables.

Since there are so many things happening in the real world, explaining why and how things happen requires creation of situations in which less is happening. Once laws have been established in simple situations, combining the effects of different variables is usually less difficult. However, even this part of the game may not be won. For example, we know how to describe the path of an object coming near a stationary object, but we still cannot solve the laws of the paths of three moving objects passing near each other.

The use of simple, controlled, artificial situations to study the relative influence of certain variables in isolation is particularly necessary in the biological and the social sciences, because the influences of extraneous variables may not be so obvious as in the physical sciences.

An example of the importance of control and artificial situations in biological sciences is the formulation of the genetic theory of heredity by Gregor Mendel. It had long been known that certain traits of one generation are passed to another. However, the explanation of this phenomenon was rather vague and was somehow attributed to the blood of the parents. This is where such expressions as "full-blooded" and "half-blooded" come from. Even plants supposedly transmitted traits to their offspring through a fluid in some manner. Both plant and animal hybrids had been grown and bred without the processes involved being understood. Certain individuals had spent many years investigating hybrids and had some knowledge about them—for example, they knew that hybrids tend to return to pure species over generations—but none had a reasonable theory to cover such phenomena.

In 1857 Mendel began to study different varieties of garden peas. He carefully studied these plants for about eight years before he presented his results. His paper furnishes an example of careful experimental control and can still be read with profit. (It is reprinted in Newman's *The World of Mathematics*.) Mendel first selected what he considered a single species for his studies. He inbred each variety for two years to make sure they were pure strains. He then limited his study by noting only single characteristics of particular plants. He artificially fertilized the plants with the pollen he wished to use, and he carefully controlled the environment of these plants so that variations in the environment were essentially eliminated. He took plants that differed in only one feature—for example, the length of the stem. He used varieties that differed considerably in length; one with a stem of 6 to 7 feet was crossed with one of ¾ feet to 1½ feet. He fertilized the tall plant with pollen from the dwarf plant and vice versa. This generation of plants—the one contributing the cross-fertilized seeds—is known as the P, or parent generation. From this set of investigations Mendel found that the question of which plant supplied the pollen and which the eggs was immaterial to the later results. This in itself was a very important discovery.

When the fertilized peas from the tall and the dwarf plants were ripe, Mendel planted them in similar environments so that he could control environmental differences. When this generation (F_1 for first filial) grew, Mendel found that all the plants were tall. There were no dwarf plants. If anything, the F_1 generation of plants was taller than either parent plant. Mendel was undismayed at the disappearance of the dwarf plants and simply continued his research.

Mendel then *self-fertilized* the plants of the F_1 generation—that is, he took the pollen of a given plant and dusted it on the eggs of that same plant. This controlled for any genetic influences except from that particular plant. When he planted the fertilized seeds of the F_1 generation, he found an interesting result. The trait that had disappeared in the previous generation reappeared; about ¼ of the new plants were dwarf and ¾ were

Figure 2—Diagram of successive generations of self-fertilized pea plants after cross-fertilization of pure strains of tall and dwarf pea plants.

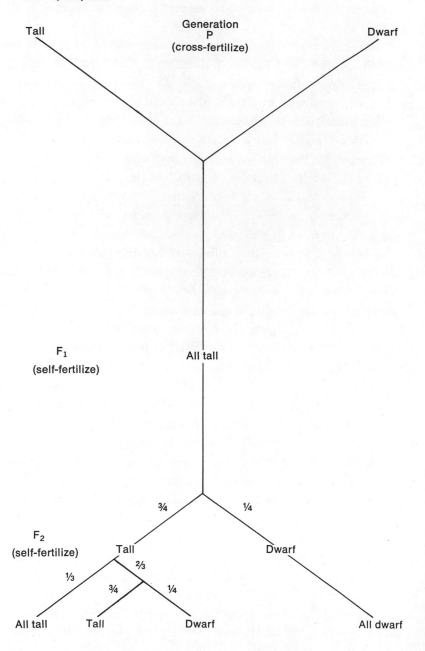

tall. Every F_1 plant had both tall and dwarf offspring. Although each F_1 plant did not have exactly ¾ tall and ¼ dwarf offspring, they all had more tall than dwarf. When the plants of this generation (F_2) were ready, they were also self-fertilized.

A description of the next generation (F_3) is quite complicated. All of the offspring of the self-fertilized dwarf plants were dwarf. Approximately ⅓ of the tall F_2 plants had only tall offspring. Mendel had no way of discriminating which tall plant would yield only tall offspring before the fact. The remaining tall F_2 plants yielded both tall and dwarf offspring; their yields were about ¾ tall and ¼ dwarf. The succeeding generations of plants that had constant offspring all had the same character as those plants. That is, if all of the offspring of self-fertilized plants were tall, all of the next generation would also be tall. In like manner all self-fertilized dwarf plants had only dwarf offspring. Based on the results of this contrived breeding system—consisting of cross-fertilizing one generation of plants and self-fertilizing all their offspring thereafter—Mendel developed interesting and important hypotheses about the nature of the transmission of parental characteristics to offspring. He identified the trait that was present in the F_1 hybrid plants as dominant and the trait that was absent as recessive. He suggested that each trait is represented by a factor; A might represent the dominant factor (tall), and a the recessive factor (dwarf). To produce a hybrid, he crossed a pure tall plant having only the factor A with a pure dwarf plant having only the factor a. Mendel claimed the hybrid had both factors represented in its cells (Aa), although the recessive factor was not apparent. The dominant and the recessive factors segregated during gamete formation—that is, each grain of pollen contained a single factor, as did each egg cell. In the self-fertilized hybrids, the factors combined according to the relation $A + 2Aa + a$. (Mendel used this formula to represent the hereditary composition of the plants. A more modern representation would be $AA + 2Aa + aa$, since the pure strains have a factor from both pollen and egg.) That is, there was one pure strain dominant and one pure strain recessive for every two hybrids in this generation.

The 1:2:1 relation came about due to random combining of grains of pollen and egg cells. Only when pollen containing the dominant character combined with an egg cell containing the dominant character did a pure dominant plant result. In like fashion, only when a pollen grain containing the recessive character combined with an egg cell containing the recessive character did a pure recessive plant result. A hybrid plant resulted from one of two ways: either from the combining of a recessive pollen grain with a dominant egg cell or from the combining of a dominant pollen grain with a recessive egg cell. From this theory, one can derive the observable results from cross-fertilizing and self-fertilizing plants. For example, self-fertilizing hybrids will yield ¼ dwarf (pure recessive) plants and ¾ tall (¼ pure dominant and ½ hybrid) plants. Only the hybrid contained both characters. In a pure strain, all grains of pollen and all egg cells carry identical factors. Therefore all fertilizations yield the pure strain.

Although the data collected by Mendel make a great deal of sense and are consistent with genetic theory as it currently exists, genetic theory is not logically derivable from the data. That is, given the theory, one can readily derive the data; but given only the data, one requires major creative inspiration to invent the theory. The relationship between theory and data is considered in Chapters 4 and 5.

One of the reasons the preceding passages were difficult to understand is that they were presented with only the information that Mendel had and not with the theory he later developed. Today it is easy to say that the same *phenotype* (the observable character of the organism) does not imply the same *genotype* (the genetic characters from which the organism develops), but for Mendel to assume that some organisms which look the same are basically different genetically was an act of true creativity. This explanation, which took Mendel eight years to develop and check, became the foundation of modern genetic theory now used to explain the inheritance of all characteristics. The explanation was not previously achieved and could not have been achieved if Mendel had not used selective and artificial means of investigation.

We have seen that scientific advancement usually depends on artificiality, selection, and control. The scientist, in order to investigate the effects of certain variables, sets up conditions where other critical variables are not allowed to have an unpredictable effect on the experimental results. By observing the behavior of objects under artificial but well-specified conditions, the scientist can formulate laws that express this behavior. Artificiality and selection, rather than detracting from understanding the events in the real world, create conditions by which one can understand that world. The real world has so many things happening at once that even geniuses such as Galileo or Mendel had to create artificial situations in order to understand natural processes.

Although selection and control are the most powerful methods for attaining knowledge about natural phenomena, there are many instances in which selection and control seem impossible. Scientists in such areas as archaeology, paleontology, political history, and astronomy cannot manipulate the situation to suit their own purposes. Their only control is to select what and how they observe. These scientists also strive to understand phenomena. They invent general principles and attempt to develop a framework within which their data is compatible. For confirmation they try to extend their predictions to include other data. They are limited in that the phenomena are always complex and they cannot always observe when and what they want. Their task in many ways is much more difficult than that of the experimental scientist, because different events in nature vary in more than one way.

Many sciences present problems that cannot be put into a formal experiment for one reason or another. With some of these problems, it is possible to use statistical techniques to isolate the effects of some of the variables. For example, statistical information confirmed the theory that cigarette smoking leads to lung cancer and certain heart conditions.

No matter what methods scientists use to isolate the effects of variables, their aim is always to achieve a broader understanding of natural processes. And it is this understanding that is the primary goal of science.

The Uncertainty of Science

Although the scientist seeks understanding, he never achieves it completely. The game of science never ends; all conclusions are tentative. No matter how much information a scientist has, he can never be certain of any of his conclusions. There are vast areas in any science where even the scientists working in that area are quite uncertain about how or why events happen the way they do. Most scientists are relatively certain that some currently accepted laws and explanations are quite accurate, but the scientists have no guarantees that they are. The problem is at base a logical one. Philosophers of science currently believe that certainty in the natural sciences is a logical impossibility. There are areas of endeavor that are certain, such as logic and mathematics, but these disciplines are certain only insofar as they say nothing about the real world.

The eighteenth-century philosopher Immanuel Kant concluded that some things about the world are known for certain—for example, that the sum of the angles of a triangle equals 180°, and that all change is continuous. But let's examine these "certainties." If a triangle is made of wood and the angles are measured as carefully as possible, the angles will not consistently total 180°. Moreover, if a large triangle is conceived as having two stars and the earth as its vertexes, and if straight lines of the sides are measured by light rays (light supposedly travels in a straight line), the angles do not always total the same. In fact, the totals vary significantly from each other. We can conclude from these facts either that light does not always travel in a straight line or that the three angles of a triangle do not always add up to 180°. In neither case are we certain that any physical entity duplicates a Euclidian triangle. We do know that the three angles of a Euclidian triangle add up to 180°; but in practice we can never be certain that we have constructed a Euclidian triangle.

Kant also "knew" for certain that all change is continuous. He meant by this that there cannot be any really sudden changes. You know, for example, that if you are traveling in a car and suddenly slam on the brakes, the car does not stop immediately; it takes about 210 feet to stop a car traveling at 60 miles per hour. If a rock falls, it gradually increases its speed. Kant assumed that by reason alone he knew that *all* change was similarly continuous. Even an explosion is supposedly a gradual, though rapid, expansion of gases. We are still not certain whether all change is continuous, but we are certain that we don't know by reason alone that all change is continuous. The question can be debated. In opposition to Kant, current physical theory states that in small systems, such as atoms and molecules, changes are in discrete units. According to this theory there is no such thing as one and a half molecules of water or five and a half photons of light. In addition, if someone turns on a flashlight, the light *immediately* travels at full speed rather than gradually increasing to maximum speed. Even electrons in their orbits around the nucleus of an atom seem to go from one orbit immediately to another, spending no time getting there. It seems just as easy to believe that all change is discrete as to believe that all change is continuous; we cannot know without experiment which is true.

And since no experiment ever provides absolute certainty (because there are always *some* uncontrollable variables), we cannot know *for certain* which is true.

We do have some certain knowledge. For example, the summation $2 + 2 = 4$ is derivable from definitions. However, we don't know for certain that 2 apples put in a basket with 2 other apples will yield 4 apples. We are pretty sure they will, but we're not absolutely certain.

This is a very difficult concept to grasp, but it is important in understanding the limits of science. There are two ways we can know with certainty that a statement is true: (1) to define it as true, or (2) to derive it logically from statements that are defined to be true. Since we can't derive any true statements about classes of events in the real world from definitions alone, no scientific statement is true with certainty. For example, we can define the concept "material object" as something that does not suddenly appear or disappear, and we can define the thing in front of us to be an apple. But we cannot then *define* the apple to be a material object; we have to *investigate* to find out whether it is one. And since we have to investigate to find out, we can only know *inductively* that the apple is a material object, and nothing that is known only by induction is known for certain, as is explained and illustrated below.

If we analyze any law or explanation in natural science, we find that at some level it ultimately rests on *induction*. In other words, at some level an assumption is made that since an event has occurred before on several occasions, under similar conditions it will happen again. One reason that no conclusions are certain in science is that there is no way of knowing *for certain* that the same thing will recur the same way. A low-level example of induction is that we expect the sun to rise every day. We may not see it because of clouds, but we expect it to be there. Why? (1) We have been told that it will rise on a regular schedule, and we may even have had the theory behind that schedule explained to us. (2) We know that it has risen every day of our lives, so why should it stop rising?

Unfortunately, these two reasons are not enough to tell us for certain that the sun will rise tomorrow. (1) Other things we have been told, and have believed, have not always turned out to be true; and furthermore, the theories accounting for the things we have been told may be wrong. (2) The evidence that an event has occurred with great regularity is not certain proof that it will continue to do so, as we can see from a short parable.

Once upon a time there lived a very intelligent turkey. He lived in a pen and was attended by a kind and thoughtful master. All of his desires were taken care of and he had nothing to do but think of the wonders and regularities of the world. He noticed some major regularities—for example, that mornings always began with the sky getting light, followed by the clop, clop, clop of his master's friendly footsteps. These in turn were always followed by the appearance of delicious food and water within his pen. Other things varied: Sometimes it rained and sometimes it snowed; sometimes it was warm and sometimes cold; but amid the variability footsteps were always followed by food. This sequence was so consistent it became the basis of his philosophy concerning the goodness of the world. One day, after more than 100 confirmations of the turkey's theories, he listened for the clop, clop, clop, heard it, and had his head chopped off. Thus, regularity does not guarantee certainty, and all induction is based on regularity.

Certainty is difficult to attain; not only can we not be certain of the same situation leading to the same results, but in actuality the "same situation" occurs only once. Assume that someone is testing the boiling point of water. He pours some distilled water into a beaker, puts a thermometer in the beaker, places the beaker in a stand, lights the Bunsen burner under it, and waits for the water to boil. After it boils he can look at the thermometer and read the marking at the top of the mercury. Now assume that he wishes to do the same thing the next day. Does he use the same water? It is now a day older. The same beaker? It has been used one more time to boil water. The thermometer also is older and has been used before. If he uses different water, a different beaker, and a different thermometer, he does not have the same situation anymore. Furthermore, it is a day later, the sun is up for either a longer or a shorter period of time, the experimenter is a day older, the earth has rotated one more time, people have been born and have died, and so on. You may feel that none of this makes a difference; but that is an *assumption* you make, as does the scientist. The consistency of the data is one way to confirm whether the assumption is essentially correct.

A scientist tries to duplicate what he deems the significant aspects of a given situation, but each situation is unique and he does not even attempt to duplicate the total situation. Even the relevant attributes of any situation cannot be duplicated exactly. The scientist accepts a certain range of values as duplication. If he is trying for a quantitative law he will try to limit the range of variation as much as possible, but it cannot be eliminated completely.

We see continued improvement in accuracy of measurement. In all of science, probably the one measure that stands as the most absolute is the speed of light. In late 1972, the accuracy of measurement of that speed was increased one hundredfold by Kenneth Evenson and his research team at the National Bureau of Standards. Whereas in early 1972 we knew the accuracy of the speed of light to about 50 meters per second, we now know it to about .5 meters per second. The current estimate is 299,792,456.2 meters per second. Even this unbelievably accurate number is not exact and will be improved upon in the future.

Another constant in an "exact" science is Avogadro's number. This is the number of molecules in one mole (gram-molecular weight) of a substance. The number is given as 6.02486×10^{23} molecules, but there is a 0.0027% error in this estimate. That is a very small error, but it comes to about 16,000,000,000,000,000,000 molecules. That is, Avogadro's number is accurate to about 16 quintillion molecules.

This lack of complete accuracy and certainty is no cause for alarm. It simply means that many of our answers can be improved. There are theoretical limits to the accuracy of our measures, but we have not reached those limits. Besides the limits of accuracy in measurement, the scientific explanations that we currently entertain are not certain to remain intact. However, it is unlikely that many scientific laws will change significantly in the near future; and whatever changes will be made will come from individuals who are committed to one of the arenas where the game of science is played. The professionals usually score the points. Successes in the game may eventually be superseded by other successes, but in spite of that possibility they are exhilarating when they occur.

In this chapter we have examined some of the rules and concepts of the game of science. Basically we found that the scientist has a systematic method that he uses to get information himself and he uses his understanding of scientific methods to evaluate the work of others. Scientific knowledge is gained solely by the use of scientific methods. These methods are never certain, but they are the best we have.

Chapter 4
Ideas and Their
Development

Where do ideas come from? They may spring forth full blown and ready for battle like Athena from the head of Zeus; but more often they appear as weak, tottering, confused babes requiring time and nourishment before they can survive independently of their parent in a fiercely competitive world.

This discussion of the origin of ideas is highly speculative, although we can consider descriptions of the origins of ideas as data. However, what is needed is not merely an accumulation of data but some way of organizing the data into a meaningful pattern. Historians, philosophers, and psychologists have analyzed creative thinking and the formation of concepts and have tried to give them a meaningful conceptual schema, but no present conceptualization can account for a significantly large percentage of the data. Despite this limitation, however, a brief consideration of the problem is worthwhile.

Our discussion is divided into three parts: first, a consideration of the kinds of people who make contributions to the game of science; second, a very brief history of science; and third, an attempt to organize at least some of the conditions surrounding the production of new ideas. Wherever possible, the discussion is restricted to new ideas in science, since even this limited area is terribly complex.

The Kinds of People Who Contribute

Today, when the amateur scientist is nearly extinct, the primary contributors of new ideas are those formally trained (or, more rarely, educated) in a particular discipline. There are at least two implications from this fact: (1) The individual has spent a very substantial period of time and effort absorbing the laws and lore of his particular discipline while achieving a graduate degree. (2) These individuals tend to be rather intelligent, at least as measured by ordinary testing methods.

Even though he has an adequate background in his discipline and a relatively high level of intelligence, the ordinary science-trained Ph.D. is still unlikely to make any substantial new contribution. Estimates vary somewhat, but a popular estimate is that about 10% of the individuals in a field contribute more than half the scientific publications. About half the Ph.D.s publish at most a single article during their career; those without Ph.D.s are unlikely to publish at all. Although published articles don't usually contain new ideas, publication is almost always necessary for communicating one. Unpublished material, no matter how original, has almost no chance to influence the direction of science.

A great many scientific innovations have been the work of young men. Gauss made contributions in his teens; Newton's important contributions started in his early twenties and Einstein's in his mid-twenties; Galileo discovered the regularity of pendulum vibrations at 17 and at 22 published work on the center of gravity of solids. There are some obvious reasons why youth has been favored. One simple reason is that there have been more young people than older ones. Some contributors died young; certainly some potential contributors did also. Another reason is that young people generally tend to be more vigorous and daring than their elders, perhaps because the young haven't seen as many "good" ideas fail. A few painful experiences do encourage caution. A third reason is that the young have not spent so many years working in a system and thus do not have the same personal investment in the status quo.

Studies of some other primate groups indicate that the old males are most resistant to change and least likely to display curiosity. For example,

Japanese monkeys were observed during the introduction of new foods and behaviors into the troop. Typically one or more young monkeys responded first to the new situation, and other young monkeys then followed. The elders, particularly the males, either were last to respond or did not participate at all. A number of studies on a variety of animals have shown that older animals explore less and display less curiosity than younger animals of the same species. There are some problems with the studies mentioned, but they are our best evidence, and studies of monkeys are in general agreement with studies of scientists. Whatever the reasons, the young tend to be innovators more often than their elders, and in spite of the primate research, men more often than women. Perhaps if it weren't for cultural restrictions, women would contribute more to science, and older women more than older men. Although few women have been active in science, Marie Sklodowska Curie—one of only two people who have two Nobel Prizes in science—was obviously a woman.

There are some personal characteristics that appear to differentiate the more creative scientists from the less creative ones. Creative scientists have a wide range of interests, which may vary over the whole range of human experience. What is sometimes reported as narrowness is quite often a lack of sociability, a limited tolerance for trivia, or interests that don't fit the usual pattern. Creative scientists may tolerate authority, but they definitely do not encourage it. Thus, they often have problems with authority figures. Creative scientists' poor rapport with administrators and other sources of authority seems to be related to characteristics variously described as ego strength, arrogance, or detachment. They seem to have a greater than average ability to tolerate ambiguity and confusion in their work, at least temporarily.

Scientific creativity is related to personality factors other than sheer intelligence. We can only speculate about these factors at present, although research on them is currently underway at the University of California and a number of other institutions.

Personal observation and various surveys of scientists agree on at least one important personality characteristic of creative scientists: they are independent. This characteristic may not extend to all parts of their lives, but in their own fields they operate as individuals. A new idea is typically the product of a single person. In order to arrive at the idea, he has to withstand pressures to think as others do. However, a team may be useful if the members stimulate one another by providing different views and information that each can then integrate for himself. Contact with other scientists for discussion is also useful in the development of an individual's ideas, since this contact forces him to formulate those ideas in some communicable form. This contrasts with the idea of "brainstorming," which was quite popular in some parts of industry. This is a technique in which people sit around and throw out ideas without criticism. Happily, this bit of nonsense has gone out of fashion, since almost all of the ideas thrown out had to be thrown out.

Since particular ideas are individually produced, clearly the creative scientist has to be independent. Pressures to conform in the realm of ideas are just as severe as they are in dress, speech, or other behaviors, and just as hard for most people to resist. Incidentally the true nonconformist is *not* one who merely thinks differently from the majority but aligns himself clearly with a sizable minority. Instead, he has his own unique viewpoint for his own unique reasons. Nudists, John Birchers, Hippies, and Theosophists are all conformists in their own way.

Historical Illustrations

In order to achieve a historical perspective, we need to abandon our view of the individual temporarily. From a distance, science is seen not as a smoothly flowing discovery of what is "out there"; rather, there are false starts, periods of stagnation, and major controversies, as well as great leaps forward. Within the major developments, there are innumerable smaller movements, countermovements, currents, and crosscurrents. Fascinating as these smaller movements may be, we shall ignore them; any detailed history requires volumes; we have only paragraphs.

To examine cultural influence, let's consider the scientific efforts of the Greeks and Romans. For two or three hundred years the Greeks lived in a period of great intellectual achievement. Science was not the major product of that period, but Greek science was surprisingly successful. Part of this success was due to major achievements in mathematics. As pointed out earlier, mathematics is not a natural science, yet it is necessary. It is an exact expression that describes the world inexactly. Mathematical acumen can be seen in the development of astronomy. Greek astronomers reached a point far beyond their predecessors, the Egyptians and Babylonians. Considering the crudity of their instruments, the Greeks' achievements were fantastic. For example, the diameter of the earth was computed to within about 50 miles of our present estimate. A celestial system very similar to that of Copernicus was hypothesized by Aristarchus and his successor Seleucus. Such hypotheses have not always been well nor widely appreciated. Plutarch reports that at least some Greeks felt Aristarchus should be indicted on a charge of impiety for supposing the earth to revolve and rotate. (Galileo, accused on similar charges almost 1,900 years later, might well have remarked "There is nothing new under the sun.") Late in the Greek era Archimedes used methods that resemble very closely those of modern science. Unfortunately Greek inventiveness largely expired with the coming of the Romans.

The great ancient empire of the Romans is an enigma. If one had to specify the conditions that should lead to scientific advancement, he might suggest a good intellectual beginning, a stable and relatively wealthy economy, places where different points of view are represented, and a common language for a large group of people. The Romans had all of these, but there is no evidence of any real achievement in science. The Romans were great engineers, soldiers, organizers, and traders; however, Roman scientific efforts and achievements as well as creative efforts in other areas were almost nil. They produced many great administrators but no great intellects. (Is there a conflict between these two roles?)

From a scientific point of view, the period between the Roman collapse and the rise of modern times is too dreary for serious comment. The beginnings of science in its modern sense can be dated from any of a wide number of different events ranging over several hundred years, with about equal accuracy. We could use the year 1210, when Aristotle's *Physics* became sufficiently well-known to be banned in Paris; or 1245, when a knowledge of Aristotle was required for a Master's degree in Paris. Some judge early contacts between Western Europe and Islam, beginning with the bloody invasion of 1097 and culminating in the quiet seduction of the Western barbarians by a combination of Greek and Arabic learning, to be the critical period for science. Leonardo da Vinci (1452–1519) can also be identified as one of the originators of science.

At any rate, by the middle of the sixteenth century, very important developments were taking place. In 1543 Vesalius and Copernicus published their major works in anatomy and astronomy, respectively. The seventeenth century produced a flood of new concepts and findings. Even though a number of historians look at the eighteenth century as scientifically dull, it seems so only when compared with the seventeenth and nineteenth centuries. Historians have generally regarded the most fruitful scientific areas in the seventeenth century as astronomy (Galileo and Kepler) and physics (Galileo and Newton); however, biology also made great strides (Harvey and Leeuwenhoek). By the nineteenth century, science was moving forward in nearly all areas.

Consideration of the history of science has brought us to a few tentative conclusions. First, there are very few examples of isolated contributions. Stimulation from others is important in the generation of ideas, and new concepts tend to be grouped in time, such as the seventeenth and nineteenth centuries, and within groups where communication is relatively easy, as in Western Europe or Greece. Second, there are some things about a particular culture and its way of thought that may either facilitate or retard scientific development. This influence has been termed the Zeitgeist, or the spirit of the times. There does seem to be some such factor operating. In effect, we have a concept that has a vague meaning, but we don't know any of its defining characteristics or exact influences.

Production of New Ideas

The third part of our examination of the origin of new ideas requires somewhat more detail than do the first two. For purposes of organization, we can arbitrarily break down the development of new concepts into four stages: (1) A problem arises in relation to an accepted concept, and that problem must be evident in some way. (2) The scientist involved has considerable knowledge in the area involved. (3) The previously accepted concept is reformulated or a new concept substituted. (4) The new concept must be developed to healthy proportions. This representation of the development of a new idea is obviously oversimplified, and these points form only the barest of bones, but they can be given some flesh through individual consideration. The points can be illustrated by aspects of the formulation of the theory of evolution.

The first stage in the development of the concept of evolution is the problem of how to account for the presence of all the diverse species of living creatures. This problem was recognized at least as early as Anaximander (about 550 B.C.). He and other Greeks raised the question of the origin of species; some entertained vague notions of evolution, while others had theories of the special creation of species. The latter theories eventually became established in Western culture primarily on religious grounds, and they provided the accepted conceptual schema that eventually broke down. As in most cases, the breakdown began when evidence appeared that seriously strained the capacity of accepted concepts. Part of the evidence was the discovery of a fantastic number of new species together with an increasing knowledge of their anatomy. The Ark slowly sank under the weight of a myriad of hooves, paws, claws, and assorted feet. The fact that skeletal structure, internal organs, sensory organs, and musculature of such diverse creatures as cats, dogs, and horses bore striking resemblances was puzzling; man also obviously had a great deal in common with many animals. Indeed, it was difficult to find any aspect of man's anatomy that did not bear close resemblance to some animal.

The geologists formulated the idea that changes in the earth's surface were orderly, not catastrophic, and took place over great periods of time. This was a major burden for special-creation theories and provided working room for an evolutionary process. It is much easier to envision an evolutionary process occurring over millions of years than over thousands. Ideas grew that strata of the earth's crust were deposited slowly, at different times, and could be dated by their fossil content. These notions not only stretched time but also cast doubt on the possibility that the number of species remained constant. Or could there have been repeated creations of species? These and other lines of evidence forced the special-creation theory into extreme contortions.

The difficulties in any accepted concept not only must be real but also must be recognized. Humans, including scientists, have a peculiar ability to blind themselves to disagreeable or uncomfortable facts or ideas. Many scientists failed to see the limitations of the special-creation theory of species as late as 1900. In particular, Germany's Virchow and France's Cuvier, each of whom was a highly regarded scientist in his own right, remained unswayed by the evidence against the special-creation theory throughout his life. In the United States the stronghold of scientific resistance to a change in the notion of the special-creation of species centered around Agassiz at Harvard. It is not completely clear whether Harvard of that day objected more to an attributed kinship to other primates or to other humans. On the balance, however, enough scientists saw great difficulties in the theory of special creation to arouse interest in and later acceptance of the theory of evolution.

The second stage in developing any new concept involves the background of the participating scientist. As previously stated, a scientist needs substantial information in a particular problem in order to formulate a meaningful new concept. This information does not necessarily lead to a new concept, of course. Science has its full share of walking encyclopedias who have a detailed knowledge about a particular area and yet leave the area totally undisturbed by new ideas. Only a few knowledgeable people develop useful new ideas.

Charles Darwin had attained a great deal of knowledge before he formed the framework for his evolutionary hypothesis. He was familiar with a considerable range of evidence from different sciences such as biology, geology, and paleontology. In short, Darwin, like other innovators, stood on the shoulders of his predecessors and contemporaries. They formed an important part of his grounding in the range of evidence surrounding the problems of species. In addition, Darwin himself was an excellent and careful observer and had the opportunity to make many new observations during his five-year voyage on the Beagle. Although formal presentation of his theory did not come for over 20 years, his background and deep immersion in the problem were critical to formulation of the new hypothesis.

The third stage in concept development has to do with the substitution of new conceptual schemas for old ones, or the reformulation of accepted ones. Although we tend to identify major changes in outlook with individuals such as Darwin or Einstein, we must remember that this is only a type of historical shorthand. Most scientific innovations are the products of many minds. In the case of the evolution hypothesis, there were many prior hints, including some reasonably well-developed systems preceding Darwin. By 1809 Lamarck produced a theory of evolution that could have served as a springboard for a more thorough investigation. Wallace had a theory of evolution that closely resembled Darwin's, and in fact Darwin's position was originally presented jointly with that of Wallace. There were a number of other scientists and natural philosophers of the early nineteenth century who had a wide variety of hypotheses about evolution.

In the "Historical Sketch" accompanying his *Origin of Species*, Darwin mentions 22 nineteenth-century writers who had evolutionary theories of sorts in various stages of development. Why then did Darwin become the principal focus in the development of the theory of evolution? What was there about Darwin's theory that led to the rather sudden demise of the theory of special creation? Why did Darwin's presentation have such a strong impact when there were so many others in the field?

An accepted theory does not collapse simply from problems; though frayed and worn, it is usually clung to as tightly as Linus's security blanket. A theory collapses only when a better alternative is available. Darwin provided a better alternative. He succeeded where the others did not, partly because the time was right, but more importantly because of his presentation. He marshalled an imposing array of data to support his reasoning. And he had a mechanism—natural selection—which was more pleasing to scientists than notions of a "vital principle" or similar mystical entities proposed by some writers. The moral of this example is that an idea, no matter how good or how timely, requires substantial supporting development and evidence before it captures the allegiance that is necessary if it is to replace an accepted concept.

Our fourth point also involves the necessity of substantial development of a new concept. Despite its impact and the fact that it was more acceptable than prior concepts, Darwin's theory of natural selection was really not adequate except as an interim solution. Had there been no further development of the theory of evolution, many scientists would probably regard it with skepticism today. A great portion of this further development has come through work in genetics. Interestingly enough, Mendel's early work in genetics was presented only a few years after *Origin of Species* was published, but those who were aware of Mendel's work at the time failed to appreciate its importance to a theory of evolution. To summarize this point, an important concept is not a static thing. New evidence and new ideas will continue to crop up; the concept must incorporate these unforeseen and often unwelcome newcomers or it will sink quietly into senility.

The Importance of Interim
or Multiple Solutions

All scientific solutions are, in a very real sense, interim solutions. The solutions we are particularly concerned with at the moment are those known to be incomplete or inadequate when they are put forth. Because there is a tendency in all humans to organize incoming information in some way, all perceived events have some kind of organization. The organization may be as simple as a collective name (such as "dog") or as complex as the General Theory of Relativity. The organization (or solution) is important because it allows us to give some meaning to the incoming information. An incomplete solution thus helps the scientist interpret some information and may point to a way to organize additional information.

Darwin recognized the incomplete character of his work in the Introduction to Origin of Species:

> No one ought to feel surprise at much remaining as yet unexplained in regard to the origin of species and varieties, if he make due allowance for our profound ignorance in regard to the mutual relations of the many beings which live around us. . . . Although much remains obscure, and will long remain obscure . . . I am convinced that natural selection has been the most important, but not the exclusive, means of modification.

How is it that the concept of evolution, admittedly leaving much data obscure or unexplained, has nevertheless been one of the most important scientific schemas to date? Darwin gave scientists assistance in two ways: a direction for research and an identification of some of the gaps in his theory. One of the most useful functions of any scientific theory is to direct research. Darwin's theory led to much important work, particularly in those areas where it was obviously weak. When Mendel's work was rediscovered it not only filled in some of the gaps in Darwin's theory but was itself given further direction by virtue of its being integrated into evolutionary theory. In brief, an interim solution may lead to research that either supports and enlarges the solution or results in a reformulation.

The world would be a tidier place if we could be sure that there is one and only one solution to a particular problem. Unfortunately, in science we are often faced with situations where more than one solution is available. When students ask "But which answer is right?" they are assuming that there is in fact a "right" answer. Scientists are more concerned with answers that organize data and are fruitful for further research. A good example of multiple problem solution concerns the nature of light. Is it a wave form, is it composed of particles, or are there other possibilities? The right answer for a scientist depends on his use of light. In some types of research it is more useful to assume that it is a wave form; in other research an assumption of particle form is more useful. Whether light is "really" a wave form or a particle goes unanswered.

A substantial number of problems in science do have multiple answers. When two or more answers seem to be about equally compatible with the data, the scientist uses convenience, simplicity, or even aesthetic values to determine the solution he accepts at a particular time.

The ambiguities that are a part of science are quite evident when we consider the possibility of interim or multiple solutions. We have identified the uncertain and ambiguous nature of science on several occasions in our discussions. Such statements are true. They do not imply that science is an inferior way to answer questions. Quite the contrary. Science, in spite of its problems, undoubtedly provides our best opportunity to resolve a wide range of questions. Recognition of its difficulties is part of the game.

Scientific Paradigms

If you open a science book in an area that you have not studied, you will notice very early in the book, probably on the first page, that the author introduces terms different from those you are acquainted with. You might conclude that the author is using jargon in order to impress you. Or you might feel that he is attempting to overwhelm you so that you won't think the book lacks substance. Most likely these are not the author's goals. Rather, the major reason everything in the text looks so foreign and forbidding is that you are venturing into a field without knowing the *paradigm* of the science.

Like most other terms introduced in this book, "scientific paradigm" is hard to define. It refers to the total complex of a science. It includes the language, theories, conceptual schemas, methods, and limits of the science. It determines which aspects of the world the scientist studies and the kinds of explanations he considers. Most important, it includes the way the scientist sees the data, laws, and theories of his science. Although a scientific paradigm contains all of these multiple elements, many paradigms are identified by one key concept.

The scientific paradigm is different for each science, but the paradigms don't usually conflict. This book deals primarily with overlapping aspects of the paradigms of the various sciences. Within the history of any science, however, there are occasions when different scientists see the world through conflicting paradigms. When this happens there is disagreement about some of the basic tenets of the science, and scientists use the weapons of argumentation, identification of relevant facts, intuition, and presentation of new data in an attempt to convince other scientists in their domain that one paradigm is better or more realistic than another. These conflicts show that the scientific paradigms are an integral part of the science.

The concept of scientific paradigms was first systematically introduced to the authors in a very informative book by Thomas Kuhn entitled *The Structure of Scientific Revolutions*, published in 1962. But precursors to the concept, as we interpret it, date back at least to Max Wertheimer in 1912. Whenever someone observes a phenomenon, he does not see raw facts, which he can then interpret; instead, he sees interpreted phenomena. There is a major difference between saying that someone sees a phenomenon and then interprets it and saying that someone sees interpreted phenomena. The latter view seems correct. Consider Figure 3. You may

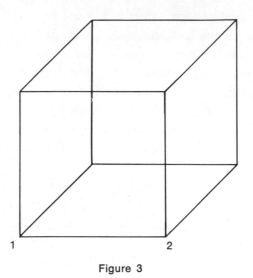

Figure 3

see a box with the corners 1 and 2 in the lower front, or you may see a box with the corners 1 and 2 in the lower back. These two ways of seeing the box conflict, because you can't see them both ways simultaneously. (If you don't yet see conflicting boxes, look at any corner; if you see it as an outside corner, gaze at it intently while thinking of it as an inside corner, or vice versa.) The important point is that you see a *box*, rather than lines on a sheet of paper as a box. You don't see the lines as two-dimensional, because you see them in different planes. You are interpreting *while* you are observing, rather than observing and *then* interpreting.

Consider Figure 4. Here you probably see six odd-shaped blacked-in areas. Now look at the figure intently. See anything? Look at the white background rather than the black boxes. Now that you see the word you may find it harder to see the black boxes. In like manner you don't look at a printed page and see merely spots on the paper, you see words of course.

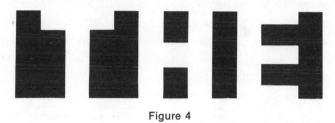

Figure 4

When you look at a page containing words in a strange alphabet, you may merely see marks on a sheet of paper. That is what this page might look like to someone who cannot read our alphabet.

Now look at Figure 5. The bottom horizontal line looks shorter, doesn't it? And the four lines on the right don't appear to be quite vertical. However, what you see and what you can measure in these cases are quite different. The horizontal lines are indeed of equal length and the vertical lines are parallel.

Figure 5

In neither Figure 4 nor Figure 5 did you actually see one thing and at the same time interpret it as something else. An interpretation is coincidental with the seeing. Of course you may later give the same phenomenon a new interpretation because you have seen them differently. But simultaneous interpretation and organization is always intrinsic to observation.

You have probably experienced new interpretations of concepts as well as of observed phenomena. As with phenomena, you see concepts somewhat differently after the new interpretation, because the observed elements fit a new conceptual schema. Consider the following series: O-T-T-F-F-S-S-E-N-T. What is the next element in the series? Think about it for a moment. . . . The letters seem to be in some sequence, but they don't all fit. An O, two Ts, two Fs, two Ss but only one E, N and T. But, now think of the numbers one to ten. Look at the series as you spell them: O-ne, T-wo, T-hree, F-our, . . . T-en. The series now fits into a conceptual schema that is different from the previous one. In one sense you now see the letters differently, and you can continue the sequence indefinitely. The letters are part of a system, and the system now accounts for the series.

To understand the series above, you have to integrate what you know from different areas: you have to know how to count; you have to know how to spell; and you have to know how to make ordered pairs between the elements of one series and a subset (the first letters) of another. Even though you know all of these things, until you see the relation between them in a new way the series makes no sense. To learn any science (in fact, to learn any area of human endeavor), you not only have to learn what the important elements of the science are, you also have to learn to see these elements in relation to the other elements. It is usually much easier to learn what the elements are than to learn to see how they are related to one another. But the learning of these relations is essential if an individual wishes to add new elements to the system or to do scientific research. Knowing the scientific paradigm is essential to extending the science, in the same way that understanding the series was necessary for its extension.

We have previously discussed two domains of science—astronomy and biology—in which the paradigm changed significantly and in which important scientific discoveries would probably not have occurred in the earlier paradigms. Consider, for example, that the discovery of the planet Neptune resulted directly from observations based on Newton's extension of the Copernican theory and that the discovery of simple fossils in Pre-Cambrian formations stemmed directly from observations based on evolutionary theory.

Once an individual learns a scientific paradigm, the world of that science becomes a different world. He sees things differently than he did before. What may have seemed central to his view before learning the paradigm may now have shifted to being merely background or second-order phenomena. As in Figure 4, when you learned to see the white space as "THE", the black marks became background. In the letter series you may have noted that some letters occur late in the alphabet and others early, but when you learned the system those became very unessential facts.

The writer of a science textbook has a very difficult problem: he knows the paradigm of the science and he is attempting to teach the science to people who do not know the paradigm. What makes the task even more difficult for the student is that many scientists, including those who write textbooks, don't realize that they see the world differently from those who do not know the science. Many scientists think that all they have to do is present the important facts, laws, and explanations, and if the student learns these he knows the science. Not so! The student, before he knows a science, must learn to see the facts, laws, and explanations in a certain way —he must learn a paradigm. The textbook writer uses a technical term to refer to a certain aspect of a class of events. He knows that what he is referring to is very important to his problem, and it is obvious to him that the technical term is central to the idea he wishes to convey. But it may be as difficult for him to realize that others don't see the importance of the ideas as it is for someone who doesn't know the paradigm to understand the importance of the term, or indeed the need for it. The scientist and the student actually see the world differently.

There is one thing that aids many textbook writers in their aims: the fact that some aspects of many scientific paradigms have become part of our culture. Although the writer may have a starting point in common with the student, the overlap is not so large that the student's knowledge can be taken for granted. It would be useful if textbook writers spelled out certain of the tenets of their paradigm in an introduction to the text.

There is one aspect of scientific paradigms that needs emphasis. The paradigm is never derivable from the data, nor even from the laws. (Most subunits of paradigms are not derivable from data either.) Scientists accept paradigms rather than derive them. The paradigm is associated with the way the scientist sees a set of phenomena. There is no necessity for him to see the phenomena that particular way. Quite often in the history of science, different scientists have looked at many of the same phenomena and have seen them differently. Furthermore—and this is an important point—no scientific paradigm with its associated theories and explanations encompasses all the data; no paradigm is completely successful. Although the scientist may see certain phenomena in a particular way, some of the data seem to make no sense when viewed that way. The scientist hopes that slight modifications of the theories are all that are necessary to make the data fit, but there are no guarantees that only slight modifications are adequate. To know the areas in which the paradigm does and does not account for the data, the scientist must know the paradigm quite well. This is a primary reason why amateurs rarely make significant contributions to the game of science today.

A discussion of some aspects of different scientific paradigms will illustrate their necessity. Consider first classical astronomy. A child may look at the heavens at night and see spots of light. He may see them as the ancients did—that is, as a large canopy over the earth with holes in it, letting through the light on the other side. He may see the sun and the moon as traveling within the canopy, floating slightly above the mountains, and giving additional light and heat. You know that this paradigm has some problems. The spots on the sky move, and they do not all move together. An individual who sees the canopy with the holes in it may worry about these *anomalies* (bits of information that don't fit). He may, as Aristotle did, add a hypothesis to his theory and decide that beneath the canopy of the sky were clear spheres housing each planet. Within this paradigm he might ask questions such as "How high is the sky?"

The modern astronomer sees the sky as a vastness beyond compare. There are in it local bodies such as the sun, moon, and planets. The sun is seen to be a huge ball of gaseous material in a constant state of activity generating a large quantity of energy. The stars are seen to be similar in kind to the sun except much farther away. The planets are solid bodies, like the earth, that travel around the sun. The clusters of stars are not organized in patterns established by the gods; they are seen in clusters only because of their position in relation to the viewer. The Milky Way is seen to be composed of millions of stars too far away to be individually identified with the naked eye. The stars are neither stationary in the sky nor moving on a canopy; each moves in its own way. Although each is moving quite rapidly, the distance between them is such that the motion is negligible in relation to the other stars.

Most of the questions that were asked about the heavens prior to this view have no meaning in the new paradigm. Everything is not accounted for here either, although an astronomer can specify the problems much better than we can. Certain problems currently exist in the paradigm for astronomy concerning the origin of planets and stars and of the universe as a whole, as well as explanations of certain visual and radio phenomena. The point is that the definitions of phenomena in the domain of astronomy have changed drastically since Aristotle's time, principally because of the new paradigm.

Most natural phenomena are part of the domain of more than one science. Scientists of the several sciences abstract parts of the phenomena differently because they view it in terms of different paradigms. The textbooks of these sciences point out the specific abstractions, but they fail to point out explicitly the paradigms related to their abstractions. Outlining a paradigm requires discussing phenomena at an entirely different level of abstraction—a level that is usually ignored.

If you peruse a textbook in physics you will find data abstracted in terms of measurement, vectors, forces, acceleration, stresses and strains, pressures, work, transverse waves, electric fields, charge, dielectrics, resistance, magnetic fields, resonance, amplification, refraction, images, lenses, diffraction, radioactivity, fission, fusion, and many others. The concepts implied by these terms are elaborated upon, experimental conditions to measure what they represent are described, and probably the relation of some of these concepts to the everyday world is discussed. But only rarely will the general conceptualization and paradigms be discussed.

Consider what might happen as you attempt to learn a new and strange language. You could memorize every word and still not be able to speak the language. You must also learn the rules, structure, and relationships within the language. The rules are learned slowly and painfully, but they eventually become automatic, giving meaning to messages in the language. Scientific paradigms are like languages in many ways, since all the elements are not immediately obvious and many steps are necessary before all the elements are tied together. If you want to realize how difficult it is to explain a paradigm, attempt to explain to someone the rules, structure, and relationships of something you know very well—for example, your native language. You will undoubtedly find many obscure aspects that you can identify but cannot express. Knowing a scientific paradigm is in a large sense like knowing a language. You can't learn a language without first learning some of the vocabulary, and in like manner you can't learn a scientific paradigm without first learning some of the specific concepts and relationships. Thus it is extremely difficult to explain the paradigm in an introductory science course.

Probably no one puts all of the concepts in any science together in one large view of the world at any given time. Scientists generally see a given set of phenomena in the context of similar phenomena within their paradigm, but they don't grasp the interrelation of all phenomena at once. An analogy can be drawn with the following complex sentence: "The mathematician that the chemist that the physicist laughed at avoided solved the problem." All the parts fit together, but one can't see them all at once. In order to understand the sentence one must analyze the different subject-object relations.

Differences among Paradigms

There are different sciences simply because each deals with different topics. The particular set of topics dealt with in a given science is known as the *domain* of that science. In some broad fashion we can say that physics started with questions pertaining to the motion of objects and their influence on one another. The paradigms of physics developed through attempts to explain how these phenomena came about. Similarly, paradigms of chemistry developed through attempts to find the essence of matter. Chemists asked "What are things made of?"; and some of the alchemists attempted to change one substance to another. Biology probably began with primitive discoveries in agriculture and medicine. Psychology had its beginnings in curiosity about the minds and behaviors of organisms.

Since early scientists were interested in different things, or different aspects of the same thing, they noticed different relationships, organized different sets of facts, created different conceptual schemas, and developed different paradigms. As we saw in the illustrations earlier, the same phenomena are actually seen differently when viewed in different contexts.

To see how the paradigms of several different sciences might be associated with a single event, let's consider a person eating spaghetti (although no scientist is likely to select this topic to study, since objects of study are also part of the paradigm and since scientists within the different sciences select only the facts that are appropriate and socially acceptable for them to study). Assume that a physicist, a chemist, a biologist, a psychologist, and an anthropologist all view this situation according to their respective paradigms.

The physicist may view the diner's fork as a rigid body that is being used as a lever. He may be interested in the problem of the spaghetti slipping off the fork, and he may note that the force of friction tends to be less than the sum of the projection of the gravitational force tangential to the plane of the fork and the inertia of the spaghetti. The chemist may note that the man is eating a starch, a homopolysaccharide that results in glucose when completely hydrolysized by acids. He may appreciate the fact that the glucose is a ready source of energy for the diner. The biologist may classify the spaghetti, stating that it comes from a particular wheat, or he may comment on the fact that swallowing is coordinated with a lifting of the tongue and that it initiates peristalsis. The psychologist might note that through experience the man has learned to manipulate the fork and spoon deftly (although he sometimes loses the spaghetti, causing measurable frustration), and that, due to conditioning, the man salivates before the spaghetti reaches his mouth. The psychologist may also notice the expansion of the pupil of the diner's eye, indicating that the man enjoys spaghetti. The anthropologist may note that spaghetti is a culturally acceptable food in certain Western subcultures and is an integral part of certain social functions (such as spaghetti dinners).

Obviously, each scientist embeds a subset of the facts into a conceptualization that his scientific subculture accepts. And each scientist emphasizes his own set of facts. This illustration does not imply that there are no similarities or overlap among paradigms. Most of this book—in fact the very existence of this book—indicates that the several sciences have many common features. But depending on the paradigm in which an event is viewed, the different facts of the event take on greater or lesser importance. No paradigm explains all the facts, and within the paradigm some facts are more important than others. Each culture, and subculture, and each person within the culture pays attention to some facts and ignores others. Some facts are more central to one view of the world than to others.

Consider yourself. You probably don't know (without looking) whether your ring finger is longer than your forefinger. You probably don't know whether you put your left sock or right sock on first or even whether you put one on first consistently. You probably do know the color of your eyes and how tall you are. You probably know the name of the book you are reading but not the names of the authors. Your age is an important fact about you if you want to drive a car, but your grade-point average is not. Your grade-point average is important if you want to win a scholarship, but whether you drive a car is not. The relative importance of facts depends on the particular situation.

If you are interested in where someone lives, knowing the house number is probably more important than knowing whether he has a large yard. Even though the city, the street, and the home number are all necessary to find the place where someone lives, the order of establishing these facts is important. The street is meaningless if the city is unknown, and the number is not very helpful if the street is unknown. An exception to this sequence may occur if you have a different paradigm for organizing the data, such as knowing that some street names have environmental restrictions. One is not likely to find a Jefferson Davis Street in New York, a William Tecumseh Sherman Street in Atlanta, or a Joseph Stalin Parkway in Washington, D.C. Moreover, one might know that a city has certain street numbers in one part of town or certain architectural styles grouped by districts.

In summary, we have seen that different people viewing the same event from different paradigms emphasize different aspects of the event; they put these aspects into different contexts; they interpret some aspects as more important than others; and they see the event differently from one another. Moreover, different paradigms lead to different organizations of data.

Scientific Revolutions

An individual scientist may puzzle over a phenomenon he does not understand; but even without understanding it, he gives it some interpretation—though perhaps the interpretation is superficial or unsatisfactory. Even such a statement as "it is out there" for a flash of light is an interpretation of some sort, because the flash could be caused by the eye or brain. The first time an event is observed, it is given a location and probably associated with other events.

When a scientist studies a set of events, he tries to fit as much of the data as he can into a conceptual schema. If he can accept the conceptual schema as part of a paradigm, he uses this paradigm as the frame of reference from which he sees the data. That is, he sees the events as interpreted by the scientific paradigm.

Usually (these days, just about always) the conceptual schema and paradigm that the fledgling scientist uses are those he learned in the textbooks and classrooms of his science. The student acquires the paradigm gradually by being told that certain phenomena go together until he finally sees the data in terms of the paradigm.

The popular assumption that science develops in an orderly, systematic fashion is orderly, wishful thinking. Such a theory assumes that the scientist goes out and gathers data, then finds by induction the laws from which the data are deduced; when a few laws have been found, they lead to a more general law. This ideal, simple sequence has been proposed as the actual process of science. But in reality science is quite different.

Scientists see facts in a framework; certain facts cannot be made to fit this framework easily. The scientist usually tries to extend his interpretation or reinterpret the facts so that they will fit it. Sometimes facts that seem to be central to the conceptual schema are not easily derivable from it. In this case the scientist usually works very hard at fitting the data to the schema—perhaps by adding new (ad hoc) assumptions or by gathering new data that he hopes will lend support to his schema. However, sometimes after much work of this type the scientist realizes that the theory on which his conceptual schema is based is unwieldy and unconvincing. The younger members of the science are usually the first to appreciate the unwieldy character of the theory. When the theory becomes unconvincing, most scientists try to modify their theory. A very few might question the conceptual schema, itself, on which the theory is based. We saw an example of this process in our discussion of the theory of special creation of species and the succeeding theory of evolution. Here the notion of what is a species and what is creation was at stake. Thus, scientists sometimes propose a new conceptual schema encompassing the data that had produced the anomaly in the old one. When there are conflicting ways of viewing the data within a single scientific domain, a revolution is in progress. New scientists entering the field may be shown both conceptual schemas, in which case they have alternative ways of viewing the data. When this has happened in the physical sciences, scientists have usually agreed after a period of time that one schema organizes the data better than the other. Either the old paradigm has continued or the new paradigm has become the way most scientists view the domain. However, in the social sciences, no one schema has ever been accepted by nearly all of the scientists. In a sense, revolutions are continually in progress in the social sciences because no paradigm dominates. Social scientists thus are faced with making a choice among conceptual schemas. Psychologists have their different paradigms, such as behaviorism and psychoanalysis; sociologists have their role theorists and institutionalists; and economists their Keynesians and classicists. It is the lack of consistent paradigms within social sciences that leads many individuals to assume that the social sciences are not sciences.

When two individuals in a given science look at the same event from different paradigms, they actually see different things, so they see different facts as being central to the science. In a sense they see past each other. Authors of introductory textbooks in the social sciences often present some of their facts within one paradigm and other facts within another. A conflicting presentation tends to cause confusion for the student, who thereby gets an inconsistent view of the field.

The fact that anomalies exist in all scientific paradigms leaves the door constantly open for new proposals. Given an anomaly, some scientists attempt to formulate a new conceptual schema in an attempt to explain it. Most scientists, however, have no desire to overthrow a paradigm with its conceptual schema; they would rather make minor modifications of a theory to account for the anomaly. In general, a view of the world that meaningfully organizes a set of data is very difficult to set aside. Because of this difficulty, most attempted revolutions by a few scientists lead to modifications of old paradigms by the majority rather than to the establishment of new paradigms. It is a rare occurrence when a revolution is successful—that is, when a paradigm shift actually takes place. But when a successful revolution does occur, it is of great importance.

Consider an important aspect of a popular paradigm of a few centuries ago. The demise of this paradigm occurred only within the last fifty years. The concept: The universe is completely filled. There is no truly empty space. If something is moved, the space that occurs is immediately filled by something else. This conceptualization was not discussed much but was a framework in which scientists and philosophers thought about and saw the world. It had been put into words with Aristotle's statement "Nature abhors a vacuum," which meant that nature does not allow open space; open space is really filled with air. Certain phenomena are derivable from such a paradigm. One of these is an explanation of how a drinking straw (or any suction pump) works. The drinker sucks the air out of the straw; since there can be no vacuum, the air is replaced by the liquid into which the straw is submerged. Undoubtedly, many individuals covered the far end of a straw, sucked, and found either that they couldn't draw anything out or else that by sucking they collapsed the straw. These facts are consistent with Aristotle's statement, but the following one was not. It was known that suction pumps could draw water only about 34 feet and had to be used in tandem to draw water above this height. This limitation is not really compatible with the filled-universe hypothesis, but no one formally questioned its implications until Galileo.

Galileo explained the limitation of the suction pump by an ad hoc hypothesis suggesting that a column of water broke at about 34 feet and therefore could not be sucked up beyond that. The same idea can be illustrated by stretching a wad of chewing gum: you reach a point where it breaks. You may have tried to stretch a drop of water over a short distance by touching it with a point and pulling it; you will notice that it is quite difficult, because the drop soon breaks. Galileo felt that water would break under its own weight at about 34 feet and therefore 34 feet was the pump's limit.

Torricelli, a student of Galileo, became interested in this phenomenon, and for some reason he didn't accept Galileo's explanations. He hypothesized a new conceptual schema in which the reason water rose in a suction pump was that we are in a sea of air and the air pushes down on the water. If the air is removed from above one spot, the air pushing down on the rest of the water would force it into the area where the air was removed. You can see the same phenomenon by putting a boat in a tub of water. The boat forces the water down beneath it and the water compensates by rising around it. Toricelli's explanation of why water would rise only 34 feet was analogous to the boat. The boat displaces only as much water as it weighs. The water that rises in the pipe rises only until its weight becomes equal to the weight of the air pressing down on the outside of the pipe. Above the raised water, Torricelli thought, we have a vacuum! He tested his theory by using mercury. Since mercury is 13.6 times as dense as water, Torricelli expected it to rise much less, only about 30 inches. He filled a tube that was longer than 30 inches with mercury and inverted it in an open bowl of mercury. The tube did not remain filled but rather the mercury fell until it was about 30 inches above the level in the bowl. Although Torricelli confirmed his expectations, others still saw the space as filled. They claimed mercury broke after 30 inches and water after 34 feet because invisible bonds could only support a certain amount of weight. Pascal, in turn, showed that the mercury would not rise as high in the tube if it were raised in the sea of air by taking it to a higher altitude, but still people believed that all space was filled. Even after Boyle showed that a liquid, which was raised in an inverted tube in a chamber, dropped as the air was pumped out of the chamber, some scientists *still* held that nature abhors a vacuum. They may have believed that Torricelli, Pascal, and Boyle's results were simply anomalies that needed to be resolved in the context of the plenum (the filled universe). According to later scientists, space was filled with an ether that was weightless, colorless, and odorless and could flow around airtight stoppers. Existence of the ether was partially based on the idea that waves need a medium to travel through, and light waves travel through space. But the advance and decline of the ether theory is another story.

The change in the view of the content of the universe from being filled to being partially empty had vast consequences. It was a successful revolution. Individuals following Torricelli, Pascal, and Boyle now saw that much of the atmosphere was open space and that weight and pressure were actually the factors in many situations in which the antagonism of nature to vacuums had been the earlier explanation. Revolutions such as this one are unpredictable in both occurrence and outcome and can be either upsetting or rewarding, but they are an intrinsic part of the game of science.

This chapter has dealt with two main themes. First, ideas do not stand in isolation. Their development depends on both personal and social circumstances. Second, ideas are held by individuals. It is the way that individuals interpret ideas and the events associated with them that constitutes a paradigm. A scientist spends many of his working hours attempting to show that the paradigm within which the events are seen does enable one to understand those events. If it does not account for important data, the scientist tries to introduce modifications so that it does. Sometimes, different scientists find themselves looking at the same data differently. If the new way of looking at the data seems to fit, a battle for the minds of the scientists takes place.

Chapter 5
Scientific
Inquiry

How is scientific information gathered? What are the relevant considerations for these operations? When is a scientific theory confirmed or negated? What are the properties of a scientific theory? What are the reasons for accepting one theory and rejecting another? These questions are basic to the process of scientific inquiry. Some of the answers have been suggested in the previous chapters. In this chapter we shall give them more attention.

In Chapter 4, scientific paradigms were illustrated, and the discussion pointed out that everyone sees the world from one point of view or another. He may switch paradigms at times, but he always operates within one. The general limits and properties of the paradigm are usually implicit in the scientific behavior of the scientist. He rarely attempts to make them explicit. In general the ideas and concepts held by a scientist seem to him to be consistent with the paradigm, even though many aspects of the paradigm are either loosely defined or not defined at all. The scientist usually considers only a small set of concepts within a paradigm and attempts to articulate, in detail, the operations and interactions of those concepts.

Scientific Theory

Within the scientific paradigm, the scientist works at several levels of specificity. The most encompassing level is the paradigm itself. The scientist has some ideas of the kinds of data that are relevant to his science and some general way of interpreting the data. He also has some ideas of the kinds of laws that are likely to relate the data and the kinds of mechanisms that are acceptable as explanatory devices. He "knows" what kind of research is acceptable and which experimental designs to use. He also "knows" which aspects of different situations are relevant and which are irrelevant. He has ways of evaluating what is good research and what is not, above and beyond the actual experimental designs employed. He will evaluate some research as appropriate and some as inappropriate, even when all the research falls within the official domain of his science. If you ask "appropriate to what?" the implicit answer is "appropriate to the paradigm." Thus the formal domain of the science is not necessarily the actual domain.

Within the paradigm and sometimes considered coextensive with it is what we have been calling the conceptual schema. This term refers to the kinds of functional relations used to describe the domain of the science. The techniques of investigation and the definition of good and bad research are not part of the conceptual schema as the term is used here, although they are part of the paradigm.

One of the most formal units within the scientific paradigm is the scientific theory. The theory is the most encompassing level within the paradigm at which any scientist is likely to concentrate his efforts.

Use of the Term "Theory"

Before continuing, we should look at the difference between the scientist's and the layman's use of the term "theory." You've heard people say "That's just theory; it's not a fact." Or "It works all right in theory but not in practice." Such statements show a great misinterpretation of the term "theory." "Theory" in science does not mean an unconfirmed statement. There may be very strongly confirmed theories as well as unconfirmed ones. In fact, a scientist is not likely to call a statement a theory until he has quite a bit of confirmation for it. He will use the term "hypothesis" or "tentative theory" until he is relatively certain that he is correct. If we wish to claim that something is unimportant because it is only a theory, then it is unimportant that the earth is round, that electricity flows through wires, and that hydrogen nuclei combine to form helium and give off energy in doing so.

It is nonsense to say that something is correct in theory but not in practice. A scientific theory is usually defined in terms of an idealized set of conditions; it may be that a particular situation outside of the laboratory to which the theory is applied is so complex that the application is tenuous, but theories are about the real world. Either they are sound in practice or they aren't sound at all. A theory is the basic formal system developed to account for observations, and a good theory is one that accounts for or explains a great number of observations simply and concisely.

More specifically, a theory is a set of interconnected statements. This set of statements comprises (1) sentences introducing terms that refer to the basic concepts of the theory (theoretical terms); (2) sentences that relate the basic concepts to each other; and (3) sentences that relate some theoretical statements to a set of possible observations. The relationships among any elements within the theory may be stated in a precise, formal way—in which case the theory is considered a formal theory—or they may be less precise, in which case the theory is an informal one.

The Relation between Theory and Experiment

The purpose of a theory is to describe and explain observable and observed events and to predict what will be observed under certain specified conditions. In order to test a theory, scientists may conduct experiments by setting up conditions for observation. It is a step toward confirmation of a theory if the events observed in the experiment were predicted by the theory, and it is a step toward negation of the theory if the events observed are contrary to those predicted by the theory. Thus an experiment provides the situation in which specific events can be observed, and the results of a successful experiment tend to confirm or negate a theory.

Sometimes a theory does not state precisely what the relations between the elements are. In fact most if not all theories are incomplete. That is, they have necessarily left at least some aspects of the relations between elements undetermined. For that reason, some experiments may be designed to provide data that will help the scientist define certain relationships among the theoretical elements. For example, in the development of theories of supersonic flight, experiments were designed to investigate the relationship between the shape of an airplane and the amount of friction in the air at and beyond the speed of sound. Prior to these experiments, theories specified the relationship only up to the speed of sound.

Experiments are sometimes designed for another purpose. Sometimes there are two or more competing theories covering the same domain of science. In that case an experiment may be designed to pit the two theories against each other. If the two theories predict different observations in the same situation, the scientist sets up this particular situation so that the observations can be made. The theory that predicted the actual observations seen tends to be confirmed and the one that did not predict the observations correctly tends to be negated.

There may be experiments, however, in which the same results are predicted by both theories. In such cases, the theories are said to be *equivalent*. If there is no *possible* way to observe events that would differentiate between the two theories—if no experiment can differentiate between them—the theories are said to be *logically* equivalent.

An example of two equivalent theories revolves around the experiment of dropping two rocks at the same time. A popular Greek theoretical explanation was that rocks and earth were made of the same element and different members of this element liked to be with each other. The theory stated that if a rock is released it goes toward the earth because it cannot bear to be away from it. As the rock gets closer, it gets so excited that it travels faster. This theory may be equivalent to Newton's theory of universal gravitation: Every particle in the universe attracts every other particle with a force directed along the line joining them, proportional in magnitude to the product of their masses and to the inverse square of their distance of separation. Some modifications of the Greek theory to include the statement that all particles like each other proportionately to their masses would make the two theories logically equivalent.

When a scientist attempts to present a new theory, he may have a difficult time getting others to pay attention to it. However, if he can design a certain kind of experiment—sometimes called an *experimentum crucis* or *crucial experiment*—he may have considerable effect on other scientists. This kind of experiment is one in which his theory makes a "psychologically unique" prediction—a prediction that is made by his theory but one that is "obviously" not consistent with the prevailing paradigmatic assumptions, and thus seemingly could not have been predicted by any other theory. Such an experiment, if successful, tends to give the theory more status than do experiments with predictions that seemingly could have been made from theories within the current paradigm. It is important to note, however, that a *psychologically* unique prediction does not imply a *logically* unique one. No prediction can be logically unique, because there are always an unlimited number of possible theories from which an event can be predicted, even if people know of only one theory. Furthermore, even theories within the prevailing paradigm can almost always be modified to account for any strange result. Psychologically unique experimental results have much more effect on the development of a science than others, although there is no logical reason for this disparity.

Consider the following example concerning the generation of lower organisms. A controversy long existed on how certain individual organisms were formed. Some scientists thought that these creatures were spontaneously generated—that they were produced by some "vital principle" or "virtue of the infusion." Other scientists thought that one living organism of a species had to produce another. Scientists on both sides did much research, but the opposition remained unconvinced. The usual experimental technique required that substances be boiled in bottles to kill the organisms in a culture. If new organisms appeared, one side argued that the bottle's cork was porous or that all the organisms had not been killed. If no new organisms appeared, the other side argued that the boiling destroyed the virtue of the infusion for generation of life or the virtue of the air for supporting life. During the early 1860s, Louis Pasteur sterilized a culture in specially designed flasks with long slender necks bent at sharp angles. Time passed but no organisms appeared. This seemed to end the controversy. It seemed that only the theory that one organism is necessary to produce another could account for these results. However, now one can easily see that the dead air in the flask could have blocked the passage of the vital principle. Incidentally, the case for spontaneous generation was reopened in 1876. In the final analysis, Pasteur's "crucial experiment," like all crucial experiments, was one experiment among many.

There is another important function of a theory as it relates to experimental research. This function has to do with the actual design of experiments. One of the functions of a good theory is that it sets up the conditions for meaningful experimentation. A person who doesn't work within the framework of a theory has no basis on which to design experiments. If you have taken a course in an experimental science, you may have wondered at the stupidity or dullness of some of the experiments. They are probably not stupid or dull to everyone. The meaning of the experiment depends on the scientific paradigm and the theories to which it relates. If you don't understand the paradigm of a science you very likely will not see the meaning of the experiment. If you understand only the methodological aspects of a theory, you can perform a designated experiment by observing specified procedures, but you aren't likely to design a meaningful experiment yourself.

Many laymen believe that an experiment must always give some sort of an answer. Once a venerable Dean, when told that a young research worker had run a series of studies but had gotten no meaningful results, asked whether scientists weren't supposed to follow the data wherever they may lead and then publish the results. But the truth of the matter is that many studies, particularly in basic research, don't give definitive answers, and may actually be uninterpretable. Most scientists have cabinets, boxes, and wastebaskets full of data that can't be interpreted. Why? There is a saying that if anything can go wrong in a study, it will—and at precisely the wrong time. The list of things that can go wrong is infinitely long: concepts may be wrong, the conceptual schema may be inappropriate, the logical sequence from the theory to the observation—which is always complex—may be incomplete, the wrong measurement may be made, equipment may falter, an assistant may prove inept, variables may not be appropriately controlled, an error in design may be overlooked, the sample being measured or tested may be atypical, there may be an unknown variable, or who knows what. In any case, the results may not make any sense.

When an experiment is interpretable, it is always interpretable within some theoretical framework. The fact that a meaningful interpretation of data requires working within the framework of a theory is missed by some of the people who write about science. They portray science as a series of blunders, a stumbling by accident from one important discovery to another. Although some discoveries do occur by accident, most do not. Scientists prepare carefully and observe carefully. Unexpected events do occur, but a scientist who chases unexpected events whenever they occur will make no progress. One reason why the scientist needs a theory to test is so that he will be able to identify those events that are predicted and those events that are not predicted as he proceeds. The careful scientist spends a great deal of time gathering and evaluating data to develop a theory that is detailed and definite enough to reveal, during an experiment, any event that is antagonistic to the theory. If he has no program of research with theories and expectations, he could not recognize the unusual. The history of science is filled with important discoveries being made and going unnoticed several times before a theory is ready to capitalize on them. If an unexpected event occurs but is unrelated to the current experimental plan, the careful scientist almost never drops everything else to follow up the unexpected event, although he is likely to keep that result in mind for possible future investigation. (Perhaps one of the reasons that accidental discoveries are played up so much is that the name given to accidental discovery, "serendipity," sounds so poetic.)

The research scientist normally performs a series of interrelated experiments. Experimental investigations are the linemen of the game of science; they don't gain much publicity but the science cannot advance without them. A single experiment cannot advance scientific knowledge very much, but it can fill a hole in a theory or help open the scientist's mind to a new insight. In any case, a good experiment will allow the experimenter to ask additional questions.

To understand some of the uses of scientific theory and to discuss in some detail experimental methodology, let's look at a detailed example.

An Example of a Scientific Theory

As our example, we can use the conditioning theory of human behavior. First of all, we should look at some of the aspects of the paradigm in which the theory exists.

This particular paradigm of human behavior developed within the scientific milieu of the nineteenth century. Scientists had begun to make predictions more precisely than they ever dreamed possible; understanding the natural world seemed within their grasp. Some laws within physics and chemistry were seen to be equivalent, and even certain aspects of living organisms were found to conform to the laws of physics. Biological scientists thought that they could reduce the complexities of life to the laws and theories of physics and chemistry. This thought received a boost with the acceptance of the theory of evolution, which proposed that higher organisms developed from lower ones. With the synthesis of organic compounds from inorganic materials, scientists could conclude, and did conclude, that all living organisms originally developed from inorganic compounds and thus that the laws of chemistry and physics could account for all life processes. This interpretation of the laws was accepted as part of the paradigm.

The paradigm, in general, has in it the belief that one should explain any complex activity as simply as possible. Thus simple laws should in turn be explained by simpler ones. In this paradigm, known technically as *reductionism*, complex behaviors such as social interaction could be explained by the laws of the behavior of single individuals; in turn, the laws of individual behavior could be explained by the laws of physiological activity, physiological activity could be explained by the chemical activity within the organism, and complex chemical activity could be explained by simple chemical action. Since the paradigm holds that simple chemical action is at base responsive to the laws of physics, its premise is, in effect, that all activity, including social interaction, is ultimately derivable from the laws of physics.

Within the behavioral sciences, particularly psychology, another application of reductionism became part of the paradigm. According to this position, complex behaviors are explainable by the laws of simple behavior. Also, the same basic laws of behavior hold for all organisms. If the laws of behavior are explained by physiological processes, it seems reasonable that the laws of behavior are generalizable among the species of animals—especially since the paradigm assumed that species were related to each other. Thus some psychologists have as part of their scientific paradigm the assumption that the same laws of behavior hold for all animals under all conditions. They assume that there may be some quantitative differences among animals and conditions but no qualitative ones. (We should note here that many psychologists accept some aspects of the conditioning paradigm as we describe it but not all aspects.)

Working within the paradigm of reductionism, the Russian physiologist Ivan Pavlov became interested in the process of digestion. Pavlov hypothesized that salivation and chewing were *reflexes* caused by food on the tongue. When food reached a certain consistency resulting from chewing and salivation, it would cause an animal to swallow; the swallowing would begin muscle activity called *peristalsis*, which carried food to the stomach. There the food would cause glands to secrete acid and enzymes, and the process would continue. According to Pavlov, one event would always lead to another until the process was completed.

To summarize the situation as it existed for Pavlov: He viewed the world through the paradigm of reductionism, believing that complex behaviors are reducible to simple ones and that they are all ultimately reducible to laws of physics. He thought that laws of digestion were the same regardless of the species and that all complex behaviors could be reduced to their component parts, which were reflexes. He demonstrated experimentally one instance of this paradigm by developing a theory of digestion. Using dogs as his experimental animals, he put them in harnesses so that he could control their behavior and operated on them so that he could observe the processes he was interested in. Carefully controlling events by stimulating the animals at different phases of the digestive process, he studied their reflexes. During this investigation Pavlov noted an unexpected phenomenon: at certain times he got erratic behavior—as, for example, when the animal salivated *before* food was put on the tongue.

Pavlov did not immediately conclude, as some would suppose, that the occurrence of salivation prior to physical stimulation proved the inadequacy of his theory of digestion. He did not drop everything to investigate this unexpected event. The accidental discovery didn't lead him to change his ways or his paradigm. He continued his work on digestive reflexes. Only after he came to some conclusion on his current line of research—enough to be awarded the Nobel Prize in 1903—did he investigate "psychic secretion," as he called anticipatory salivation.

Since Pavlov's paradigm included the concept of reflex behavior, he decided to find out what event led to psychic secretion. Pavlov reasoned that if the animal salivated before receiving food because of a reflex, then some event prior to the salivation must have initiated that response. Thus salivation must have been caused by some change in the animal's environment.

Before continuing with the discussion of Pavlov's research, we need to digress for a bit in order to consider the concepts of variables and experimental design. These concepts have been discussed on a common-sense level earlier in the book, but the specific kinds of variables Pavlov used and the conditions under which they were observed are more easily understood from a more formal point of view.

Variables

Almost all discussions of experimental design and scientific methods either explicitly use the term "variable" or imply the concept of variables. A variable refers to a specific aspect of events or things. It differs according to the particular event being considered, and no single event or thing can have more than one value of the same variable at a given time. The concept can best be understood through examples. The weight of a person is a variable: different people weigh different amounts, and the same person weighs only one amount at any given time. The number of times you blink in an hour is also a variable; in any given hour you will blink a certain number of times and this number may differ from hour to hour.

Variables are not necessarily personal. The color of cars, the make of cars, the number of cylinders in a car, how far the car goes on a gallon of gasoline, how much the car cost—all are variables. The concept represented by the term "variable" is a very pervasive one. Whenever an entity or event is described, every dimension by which it could be described can be considered a variable, including where it is and its environmental conditions.

One of the most important things a scientist does when he conducts an experiment is to attempt to find out what the relationship is among different variables.

Let's consider what is meant by a relationship between variables. How fast can you run 100 yards? It is obvious from what has been said that the time it takes to run 100 yards is a variable. It is a variable that can be quantified. To be quantified simply means that the different values the variable takes are established by some method of measurement. Quantitative values are usually numbers. But how fast can you run 100 yards? What if you were running on grass? What if you were running uphill? Or after a large Thanksgiving dinner? What if you were wearing hip boots? What if it were raining? Such things as the kind of track, the slope of the ground, the time since you last ate, the kind of clothes that you are wearing, and the weather conditions are all variables. They all probably have some effect on your running speed. The color of the clothes you are wearing, the depth of solid rock beneath the track, the time since you last went to the movies, and the number of sisters you have are also variables. These latter variables are unlikely to have much effect on running speed; there is a relationship, however, between the kind of clothes you are wearing and running speed, because the heavier your clothes the slower you run. Race horse handicappers use this relationship between variables in an attempt to equalize a horse race; the horses that are known to run faster have to carry more weight.

If it is true that the more weight a person carries the slower he runs, we can say that there is a *functional relation* between weight carried and running speed, or we can say that running speed is a function of weight carried. If we wished to state this relationship in symbolic terms we might do it this way: let S equal running speed and W equal weight carried; then $S = f(W)$. This equation is read as "S equals f of W," or "S is a function of W," or "speed is a function of weight carried." A more technical way to state this part of experimentation is to say that a scientist investigates the functional relationships between variables. He studies the systematic changes in one variable as another variable changes. The scientist may simply be interested in establishing the general nature of the functional relation, such as "the more weight one carries the slower he runs." If he is interested in establishing a specific relationship between variables, he might do so by constructing a verbal description, a graphic illustration, or a mathematical equation.

Experimental Design

What are the problems encountered when a scientist conducts an experiment? In designing the experiment he has to be very careful to see that the relationships he finds do exist between the variables he is investigating. He must try to hold *constant* the influence of variables other than those he is investigating.

In an experiment, there are three kinds of variables: (1) independent variables, (2) dependent variables, and (3) control variables.

Independent variables. Independent variables are those variables whose values are directly manipulated by the experimenter. The experimenter is interested in establishing functional relationships between independent and dependent variables. If, for example, an experimenter is interested in determining the relationship between weight carried and running speed, he can do so by getting someone to run 100 yards on a number of different occasions. The runner can carry a pack each time he runs, and the experimenter can add some rocks each time. The runner may run six times; he can carry either 0, 10, 20, 30, 40, or 50 pounds of weight each time he runs. Since the experimenter can directly manipulate the amount of weight carried, weight carried is an independent variable.

Dependent variables. Dependent variables are measures of events that occur during the experiment or are measures of conditions existing after the experiment is completed. Whereas the values of the independent variable are in a sense assigned before the experiment begins, that control is not possible with dependent variables. Dependent variables are those that are affected by the actual process of the experiment. They are variables whose specific values, in principle, cannot be measured before the experiment begins. In our current example the dependent variable is the time it takes the runner to run 100 yards. This time is measured under each of the values of the independent variable. If the experimenter wants to state the results in the form of a functional relation, he says that running speed (the dependent variable) is a function of weight carried (the independent variable).

Control variables. Control variables are crucial. They are the conditions that make an experiment an experiment. These are variables that the experimenter does not want to vary systematically with the independent variable. In other words, the average value of the control variable should not change as the independent variable changes. The experimenter thus sets up experiments so that these control variables cannot systematically affect the relationship between the independent and dependent variables. The major problems in experimentation are associated with control variables. Consider our example. If we are investigating the relationship between running speed and weight carried we do not want other variables to influence running speed. We would want the race to be run under as similar conditions as possible for each of the values of the independent variable. The runner should be well trained before the experiment begins so that he does not improve with practice during these six occasions. He should run each race on successive days so that his energy will not decrease with each race. He should run all the races alone, or all of them with someone else, so that the desire to compete will be about the same each time. He should run each time on the same track so that the composition of the track will not affect the outcome. He should run each race at the same time of day (allowing the same amount of time to elapse after eating), so that the weight of the food and the physiology of digestion will not have different effects on the outcome of the race.

In fact, to be more sure of the relation between running speed and weight carried, the runner should probably carry each weight more than once. There are so many possible influences on the time it takes to run that the experimenter cannot possibly control them all. Say the runner runs with the same six weights on many occasions. He runs once a day for three months, with the conditions as nearly the same every day as the experimenter can manage except for the weight carried. On successive days he carries different weights, and he carries the same weights regularly throughout the period; then different influences—such as changes in humidity and temperature, which the experimenter may not be able to control, and the feelings and attitudes and temporary physical conditions of the runner, which the experimenter certainly cannot control—will tend to average out over the different values of the independent variable so that they will not systematically influence the final result.

Another way to attempt to control the influence of other variables in finding the functional relation between weight carried and running speed is to give to each of a number of different people the same weight to carry and have each run 100 yards, and then to give other people different amounts to carry, until many people have run with each weight. In this way the influence of all the other variables will again tend to average out. Some people will be more tired than others, some in better shape than others, some more highly motivated than others, but hopefully some of all kinds will run in all conditions. The results under these conditions will not be very precise for any individual, but they may be precise enough to show a relationship between the independent and the dependent variable, and the experiment will take less time than the first method.

To be sure we understand the important aspects of basic experimental design, let's briefly consider a few other hypothetical experiments.

We might be interested in the relationship between frequency of vibration and tension on a wire. The independent variable is the tension. We manipulate tension by attaching the end of a wire to a hook, stretching the wire out horizontally, running it over a pulley, and attaching this end to different weights. We can attach 10, 20, 30, 40, or 50 pounds. The amount of weight is the independent variable and an operational definition of tension. The dependent variable may be how fast the wire vibrates. We can estimate vibration by plucking the wire and determining its pitch, or we can attach a very lightweight pen to the middle of the wire and allow it to touch a sheet of paper moving at a constant speed under the pen. The picture drawn will be a sine wave—that is, it will look somewhat like this: ∩∪∪ . The closer the waves are together, the faster the wire is vibrating. The control variables may be how far the wire is pulled when it is plucked (it should be pulled the same amount for each tension), the material of the wire, the diameter of the wire, the humidity and temperature of the room, and so on. Each of these should be the same or should vary the same amount for each tension.

Another example is the relation between the amount of a substance dissolved in water and its boiling point. The independent variable here is the amount of a substance dissolved—such as 1, 2, 3, 4, 5, or 6 teaspoons of salt. The dependent variable is the temperature of the water when it boils. Control variables are such things as the amount of water, the consistency of the salt being dissolved (the same type of salt must be used for all amounts), the kind of container, the temperature and humidity of the room, and so on.

In summary, an experiment contains three primary elements:

1. Independent variables, which are directly manipulated by the experimenter so that each experimental condition has a particular value of each.
2. Dependent variables, which are measures taken during the experimental process.
3. Control variables, which are variables that should *not* vary systematically from condition to condition.

The experimental scientist has the tasks of trying to relate some terms of a theory to dependent variables and some of them to independent variables. Then, if he can design an experiment so that he can observe under controlled conditions, he will find out what the relationship is between his independent and dependent variables and, through them, what the relationship is between his theoretical terms. If the theory makes predictions, the experiment can tend to confirm or negate the theory. If the theory is not explicit, an experiment can give the evidence necessary to make it more explicit.

An Example of the Relation
between Experiment and Theory

Now let's return to Pavlov and the example of conditioning. Pavlov was interested in finding out how events that preceded food on the tongue came to elicit salivation. He thought that such erratic behavior was due to reflexes of the higher nervous system. Pavlov thought he could find the conditions that caused this erratic performance. To investigate it, he first had to define a dependent variable and an independent variable. Salivation was an obvious dependent variable. Since Pavlov believed that some event preceding it led to the salivation, he varied the presentation of a cue before salivation. His independent variable was whether or not a metronome sounded before food was placed on the tongue. Many controls were necessary to do the experiment. Pavlov used dogs as a control variable so that he would not obtain variable effects from different species. He put the dogs in a harness and in a room with extraneous variables held constant. For example, the room was kept at a constant temperature and humidity. It was both soundproof and lightproof. Food was delivered directly to the mouths of all animals without any physical disturbances in the room. The dogs were in the same state of hunger for all conditions.

Pavlov found that the first time the clicking sound of the metronome came on, the dogs perked up their ears and looked toward the sound; then they received the food, which caused salivation. The same sequence of events was repeated. After a very few pairings of the click and the food, some of the animals started salivating before the food was presented. Pavlov had rediscovered psychic secretions; this time, however, they did not seem mysterious at all. Now he knew that the salivation *followed* the click, even though it preceded the food. Salivation was a response to the click.

After the original experiments Pavlov varied the interval between onset of the metronome and presentation of food. Pavlov found that salivation was more likely to be elicited by the click if the click began five or fewer seconds before the food was introduced. Amount of salivation was a function of the time between the onset of the click and food presentation. Pavlov had demonstrated that the complex behavior of getting ready to receive food could be explained by procedures he was familiar with. In fact, it was no more than a special kind of reflex. This reflex, Pavlov thought, was due to the action of the higher brain. Pavlov called salivation due to the click a "conditional reflex" because it was conditional to (or dependent upon) a specific experience—the pairings of the click and the food. Salivation due to the food itself was an "unconditional reflex" because it did not seem to depend on the animal's previous experience. The terms *conditioned* and *unconditioned* are now generally used instead of "conditional" and "unconditional."

Pavlov found in later investigations that any "neutral stimulus" (such as clicks, lights, or buzzers) could be associated with an unconditioned reflex and then elicit the response of that reflex. Conditioned reflexes can thus be established by pairing neutral stimuli with unconditioned reflexes. The conditioned reflexes tend to disappear when the neutral stimuli are presented alone. Many other laws were developed, and conditioning, as the technique came to be called, became one of the fundamental procedures in the study of learning. The techniques and results would not have been found if Pavlov had not formulated a theory with which to interpret the events.

More recent research in animal learning has led to a great number of specific additions and changes to the theory. Some psychologists think that the research has led to modifications that are awkward, perhaps contradictory, and untenable. If the theory is rejected it will be because detailed experimentation on anomalies has shown its inadequacy and has led to a body of data that can be interpreted more simply through a different theory. Thus, even a faulty theory serves a major purpose: without it, better theories cannot be developed.

Formal Characteristics of Science

We have discussed some of the factors that are involved in experimentation, and we have seen that experimental results provide data that is used by a scientist to confirm, develop, or modify a theory. Now we should briefly discuss some formal concepts involved in the confirmation and development of theories.

A scientific theory can be viewed as a set of statements about a certain domain of the natural world. The theory is an attempt to describe the activity of the events in that domain. Some events are easily observed, while others can be observed only indirectly and by devious means. Most scientists believe that some events are difficult, or even impossible, to observe. They justify such a belief by assuming that consequences of these events can be observed, and that the consequences can be explained better by assuming that unobserved events occurred. A scientific theory is an attempt to represent, in some symbolic form, unobservable events as well as those that can be observed. Most theories use the symbolism of a natural language (such as English) in combination with mathematics, although some stay entirely within the realm of natural language. A recent development—one that has some unexplored consequences—is to use the symbolism of a computer program to represent the activites of natural events.

Whatever the form or language of the symbolism, there are a few basic concepts implicit within it. In a sense a scientific theory begins with the naming of a set of theoretical entities. Then there is a set of sentences that define functional relations among the entities. There may also be another set of sentences constructed out of the original set to define new entities. Finally, there is still another set of sentences that relate some of the theoretical entities and their functions to possible observations.

A theory tends to be confirmed if actual observations agree with those predicted by the theory. If actual observations are inconsistent with those predicted, there is an error somewhere in the system, although an inconsistency does not show exactly where the error exists. There are occasions when the theory does not specify exactly what the observations ought to be; in that case the data collected are used to help specify the functional relations within the theory.

Let's consider some of the different aspects of scientific theory. First, a theory is a set of sentences, some of which may be expressed symbolically. The theory is found in books and journals, it is not found in nature. True, the theory is supposed to refer to events in the natural world, but the theory itself can be analyzed at one level in terms of its symbolic form.

The theory formally begins with a set of undefined terms that name the basic theoretical entities. They are called "undefined" because they are not defined by other terms. The fact that these terms are formally undefined, undefined in the syntax of the theory, does not mean that the terms have no meaning; we can conceptualize the terms, imagine them, and perhaps even see them without changing the possibility that they are undefined. It is a logical necessity within any symbolic system, if circular definitions are to be avoided, to have undefined terms. Other terms are defined on the basis of the undefined terms. Consider an unabridged dictionary; theoretically, every term in the language is defined in it. But every word in the dictionary is defined using other words. If the words in a dictionary were your sole source of meaning for every word, no word would have any independent meaning since each word's meaning would depend on the meanings of words you did not know. (Consider studying a dictionary in a language you don't know. Reading the meanings doesn't help.) The scientific theory takes what seem to be the basic theoretical terms and accepts them without defining them. A few examples of un-defined terms are electron, negative charge, meter, second, gram, element, reinforcement, and habit. These are the basic terms of certain theories.

Second, a theory contains a set of sentences that defines relations be-tween the theoretical terms and defines new terms using the undefined terms. The relations between terms may be stated in either a very formal language or relatively informally. One of the uses of mathematics in science is in specifying the functional relations between theoretical terms. An example is the relation between time and the distance traveled from rest by a falling body; in Newtonian physics this relation is specified as $S = \frac{1}{2}gt^2$. Another example is the relation between a stimulus (R) and a sensation (S) in Fechner's psychophysics: $S = K \log R$. In more informal theories, instead of the formal mathematical relations between theoretical terms, the relationships may be very general or even unspecified. In political science, for example, there is a statement that the more personal the means of com-munication the more effective it is in changing opinions. Another example is that good congressional politics is not always good presidential politics.

Another kind of sentence in a theory is that which relates certain aspects of the theory to possible observations. These sentences are given different names. Sometimes they are called *coordinating definitions,* sometimes *operational definitions,* and sometimes *reduction sentences.* Whichever they are called, these sentences form a vital part of the theory. They take a formal exercise and turn it into an empirical science. If there is no way to relate the theory to observations in a consistent manner, then no matter how good the theory sounds it cannot be considered a meaningful scientific theory. An example of a coordinating definition in nuclear physics is: "An alpha particle when viewed in a lighted cloud chamber is seen as a short, relatively thick white line; a beta particle is seen as a longer and thinner line." An example from psychology is: "Intelligence is measured by the score on an intelligence test." From economics we have: "The stock market level is identified with the Dow-Jones average." In addition to expressing formal relations between theoretical entities, another use of mathematics in scientific investigation is through statistics, which compare how closely the observations agree with those predicted in the coordinating definitions. The actual measurements are not exact, and other uncontrolled variables may affect the observations somewhat. Because of this inexactness, the observation (or the *average* observation in a number of experiments) is expected to be within some specified range of the predicted observation. Identifying the range of acceptability is a function of statistics. For example, Kepler was able to assume that his theories were confirmed when he observed a planet within about four minutes of arc of where he predicted it would be; his coordinating definitions had specified within limits the direction and the height of a spot of light.

Thus a scientific theory formally consists of (1) a set of undefined terms, (2) a set of relations between the terms, (3) new terms and laws defined through the undefined terms, and (4) a set of coordinating definitions relating some of the sentences in the theory to possible observations. The process of scientific inquiry is an attempt to relate actual observations to those possible observations predicted by the theory.

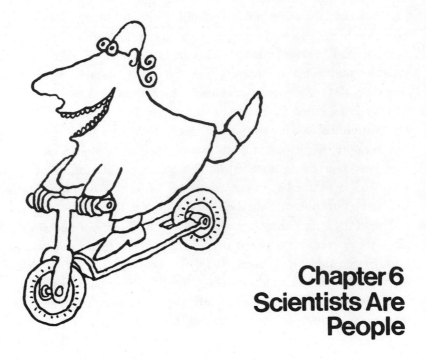

Chapter 6
Scientists Are
People

Previous chapters have presented various aspects of the game of science. But we have not as yet dealt directly with an essential component of the game—the participants. We have talked about scientists from time to time, but our principal concern has been with science itself rather than with the people involved in it. In this chapter we'll take a look at the scientists—their interests, motivations, values, status systems, rules, activities, and education. A large part of the chapter is necessarily based on the authors' personal observation and conjecture.

Motivations and Interests of Scientists

At least two groups who work as scientists can be identified according to their motivational systems: first, those motivated by the intrinsic pleasures of playing the game (the Players); second, those motivated by a desire for recognition and its resulting rewards (the Operators). These two groups are not completely exclusive, but they are worth examining separately. In addition, because individuals are complex, it is doubtful that either of these systems of motivation can account for all the drives that lead any scientist to the intense and prolonged effort necessary to produce a significant contribution.

There are other people trained as scientists who either fail to function as scientists at all or on completion of their formal training no longer attempt to add to the body of scientific knowledge. Their professional motivations are at least partially different from the first two groups. There are quite a few members of these last groups, so although they contribute very little to science directly, they should be considered.

The Players

Our first concern is that happy breed who find the game endlessly fascinating. For the Players the reason for the game is not primarily knowledge or the good of humanity, but simply the game itself.

For purposes of organization we can arbitrarily break down the Players' motivations into six topics—which incidentally include most of the attractions of games listed in Chapter 1.

1. Curiosity.
2. The delights of ambiguity and uncertainty.
3. The contest with nature.
4. Escape from the boredom and crassness of everyday experience.
5. Aesthetic pleasure.
6. The sheer joy that comes from exercising the intellect.

1. *Curiosity.* The first two topics, curiosity and ambiguity, overlap in some ways, but it seems worthwhile to separate them. Curiosity (sometimes equated with the tendency to explore) is a very basic impulse for all higher vertebrates and is also present in many creatures lower in the phylogenetic scale. Such tendencies seem to be primarily innate in humans. Curiosity may sometimes kill cats, but it is more likely to produce new opportunities for learning, location of food, and contacts with suitable mates. There is experimental evidence that more complex species tend to display a higher degree of curiosity than do less complex species. Other evidence, ranging from children rattling and squeezing packages to starving animals exploring a new environment before eating, demonstrate that curiosity is indeed a very strong source of motivation.

Scientists have very strong curiosity impulses, and their curiosity is somewhat different from the curiosity of most others. Two differences are clear. (1) The scientist's curiosity about a particular problem may continue through his entire career. (2) The scientist's curiosity in his field is generally impersonal.

Within the scientific community, curiosity plays a particularly large role. For people who have a lively sense of curiosity, science can provide never ending opportunities to exercise it. The quest for the unknown is as basic to the scientist as it is to the explorer, the philosopher, or the artist.

2. *The delights of ambiguity and uncertainty.* Scientists not only tolerate ambiguity and uncertainty, they often seek them. Ambiguity generates tension, and for some this tension can be extremely rewarding, even though it continues over a long period. For others, continued ambiguity-produced tension may be shattering. The experience of tension is related to un-resolved or incomplete situations. For example, consider a professional football game. What makes a game exciting? Isn't it the tension and excitement due to an uncertain outcome? Why? Which is more interesting and exciting, a contest won or lost in the last 14 seconds or one in which a team wins by a large margin? The uncertain outcome is extremely impor-tant. When we watch the TV rerun of a game, our excitement and interest is much less than that of a live game. The tension resulting from uncer-tainty plays a large part in our enjoyment. Although science is different in at least some respects from professional football, science also generates the tension of uncertainty. The scientist's informal description of routine projects as "scut-work" or "hack-work" gives the flavor of his feeling about activities in which the outcome does not contain sufficient uncertainty to evoke tension.

Scientists don't ordinarily seek ambiguity or uncertainty in all phases of their lives. They want their cars to start regularly, their grant requests approved without delay or change, and other routine facets of their lives to proceed without undue confusion or surprise. In general, their enjoyment of uncertainty lies in the challenge of unresolved problems. Even here the enjoyment of uncertainty and ambiguity has its peculiarities. Scientists are motivated to reduce ambiguity and uncertainty, but they are stuck with the realization that a reduction of uncertainty in one sphere almost always brings about the awareness of new problems with its uncertainty and ambiguity to be resolved. To summarize, the tension resulting from ambiguity and uncertainty is among the important motivational factors in science, and science can be the most uncertain of games.

3. *The contest with nature.* The old saw of climbing mountains because they are there is illustrative of the fact that a contest with nature can evoke a strenuous response. The principle is very much the same in science, even if the response is ordinarily not so strenuous. Consider the career of a great man and an excellent scientist, Jean Henri Fabre. He spent a great portion of his life studying the behavior of insects. Between the ages of 74 and 92 he worked almost continuously on such observation. This period is of particular interest. What drives a man in his eighties on when he is no longer employed in an academic position and has long since achieved fame through his studies? Fabre appears to have taken understanding the insect world as his own peculiar contest and would not rest so long as he was able to continue functioning. There has to be a very strong quality of attraction that leads a man of 80+ to squat for eight hours watching a solitary spider spin its web. Many amusements lose their savor as we age. In this game lies the possibility of perpetual youthful savor if not perpetual youth.

Incidentally, Fabre's fascinating works are easily readable by the layman. One imagines that, like Bertrand Russell, Fabre felt secure enough in his knowledge and position so that he could write clearly, unencumbered by the jargon of his trade.

4. *Escape from boredom.* How do you escape the sleazy world of TV hucksters, the mundane world of everyday life, and the anxieties generated by personal problems? Some turn to murder mysteries. Here they can substitute their own favorite character for the victim. Other people turn to drink, social events, sex, travel, or Sunday painting. The scientist has a built-in escape in his work. When he is involved in his scientific work, other problems vanish. Einstein's address at the Physical Society in Berlin in honor of Max Planck's sixtieth birthday focused on this particular point: "I believe with Schopenhauer that one of the strongest motives that lead men to art and science is escape from everyday life with its painful crudity and hopeless dreariness. . . ." The point couldn't be stated more succinctly.

5. *Aesthetic pleasure.* Many assume that science and the diverse arts such as poetry, music, sculpture, and painting are at opposite ends of a

continuum. They are wrong. Like many other human endeavors, science has its aesthetic side. Some theoretical positions or problem solutions are preferred because of their elegance. There are beautiful problems and beautiful solutions. Aesthetic charm in science may rest heavily on lean, simplified lines that are both functional and soaring. Artists and writers differ from scientists in the language they use and in the material they appreciate, but the aesthetic attractions of their game are remarkably like those of the game of science.

The excitement of a strange and lovely object is a form of aesthetic experience. The explorer setting foot on an unexplored planet or the artist stepping back to view his painting must experience some of the same feelings as the scientist who makes an original observation or creates a new idea. It is the experience of oneself as a completely unique individual. Small wonder that Archimedes leaped naked from his bath shouting "Eureka!" when he discovered his principle. At that time only he and the idea existed.

6. *Joy of exercising the intellect.* For an athlete, exercise of his prowess can be a thing of joy. The professional athlete has a pride of accomplishment, and part of his reward is the opportunity to display his capability to his admired peers. The same sort of thing appears true of the artist, ballet dancer, or actor. In each case, however, it is the exercise of capabilities for one's own benefit, plus the approval of a few peers, that is important. Scientists don't differ greatly from other professionals. They enjoy the exercise of their intellectual capability in the same way athletes enjoy exercise of their physical capabilities.

There is also a desire for approval by their admired peers. Mere public acclaim may be considered bad taste or even an indication of foolishness, whereas one word of praise from a particular colleague is an event of singular importance. Overall, the motivations of the scientist resemble those of many other professionals; they include the joy of the game itself, together with self-esteem and the approval of valued colleagues.

The Operators

The motivations and goals of the Operators are primarily recognition and its accompanying rewards. (There are good examples of this motivation in fictional characters such as Milo Minderbinder in *Catch 22*.) The

Operators have some of the same characteristics as the Players, but they differ greatly in their objectives. This statement is not meant to imply that the Operators contribute nothing to science or that they are unneeded.

The Operators seem, on the average, less gifted intellectually than the Players. In spite of these lesser intellectual gifts, however, the Operators are often quite successful. Operators tend to be very active; in fact, their activity is ordinarily more visible than that of their colleagues. Many Operators get large grants or contracts to do research and capitalize on the contributions of their junior colleagues and students as well as their own. Operators are particularly noticeable in the social realm, serving as working officers of scientific organizations, for which they organize meetings and perform administrative tasks. Many Players do such jobs out of a sense of obligation, and sometimes, but rarely, because they enjoy or aspire to these jobs. Acceptance of an administrative position makes serious scientific work difficult or impossible. A few scientists continue to do scientific work in spite of an administrative job, but most cannot. Thus Players are rarely interested in even discussing administrative positions unless they feel an obligation to do so.

Since the Players are under-represented, those who occupy positions as administrators or organization officers often present a distorted view of their discipline to the outside world. These are the scientists who have a greater likelihood of coming into contact with laymen. Many Operators are found in industry, where the pay and other external benefits are higher and the odd wanderings of the self-contained scientist are not acceptable. A combination of rewards is sufficient to attract most Operators to nonacademic pursuits.

A single distinction between Players and Operators might be that the Players are primarily internally motivated while the Operators are primarily motivated by the external rewards.

The Coaches

There are those who use science as a means toward an end. They enter the world of science as trainees because they think the game is an interesting one to study, but they do not directly play the game themselves. They may prefer to disseminate scientific knowledge to others as a teacher, textbook writer, or reporter. Some people are very skilled at coaching, or science writing, rather than playing the game. These are very important functions and many Players like to think that one must be a Player to do them well. A diligent Coach may give a potential future Player a vision of what the game is like if he can deliver the feeling of the game's excitement and explain the latest versions of some of its fields. It is tough for a non-Player to do this, however, because much of the excitement comes from the dynamics of the game's activity, which in most instances must be witnessed to be appreciated. Thus, for a Coach to become successful requires continuing hard work. The final preparation for becoming a Player, however, should be at the field of the game; a graduate student almost always must participate as an apprentice before he can become a full-fledged Player; the final training of successful Players is almost always with someone who is, himself, playing the game.

The Bystanders

The fourth group—those trained as scientists who never play the role of scientists (the Bystanders)—is puzzling. What gives a man the drive to complete a long and trying period of graduate study and then terminate his direct involvement when his degree has been completed? The primary factors are probably personal ones, stemming from a lack of independence or self-confidence. Given the uncertainties of the game, a high degree of confidence, or even arrogance, is required to function outside the protective armor of a graduate school. Consider the situation of a young scientist who is not yet established. He has been taught that research, to be of value, must be original. But to be original, he must defy the "knowledge" of his established elders and point to their errors. This is no position for an intellectual shrinking violet. David had only one giant. The young scientist who advances a new idea arouses an army of giants. In this army are not only foes but his intellectual fathers as well. The odds in such an encounter are heavily in favor of the giants. Many young scientists opt against originality and, although they remain scientists, do mundane work. Many others trained as scientists find a quiet niche and view the carnage from afar with singular detachment. The latter two alternatives are rational and thoroughly justified. Only madmen would do otherwise. Fortunately we still have our madmen.

There are at least three more possible reasons for individuals becoming Bystanders. One of these is good old-fashioned laziness. The newly graduated Ph.D. has worked under pressure for several years. There is a strong temptation to relax. Once relaxed, the creature may send down roots, and become satisfied and comfortable in his present state. Recovery from that state is likely to be delayed indefinitely.

The second reason for terminating serious scientific effort may be a shift in interest to nonscientific pursuits. The shift may be to some activity completely outside the intellectual realm, such as physical culture, Boy Scouting, or the maintenance of the ideal middle-class home. Shifts to activities inside the intellectual community, such as art, philosophy, or history, are relatively easy to understand. Of course a scientist interested in art, exercise, or his home is not necessarily a Bystander. The critical factor is whether he continues to contribute to the body of scientific knowledge.

Third, the reason simply may be boredom. Unfortunately, many who have tried playing any game day in and day out for a long period find that what starts as a delightful pastime may gradually become boring and end as a loathsome task.

Scientific Theorists

Let's now briefly consider a generally honored but sometimes maligned type of scientist, the theorist. Theorists are considered separately, since they are the rarest, most fascinating, and most important of the species scientificus. Their motivational system most often is that of the Player, although occasionally it is that of the Operator.

Although theorists are often viewed as cold, rational, deliberate machines, they are generally almost the opposite of this popular picture. They are usually men of strong feelings who have the ego of actors and an irrational, almost mystic attachment to particular views of their discipline. The appearance of cool deliberation is their public face, which often represents only their disdain for contact with the spectators.

There are important occupational differences between theorists and other scientists. Theorists set the framework within which others do their research. Those other than the theorists do the important work of filling in details of existing theories. Non-theorists fulfill a relatively safe and useful function. Their work contributes to science but does not threaten the individual scientist unless he happens to accumulate evidence contrary to the status quo.

What is the general personality makeup of theorists? Are they normal, neurotic, or even psychotic? They rarely fit the pattern of middle-class normality, and yet they are intensely in touch with their own reality. Perhaps they don't fit any of the usual categories. Just assume that they are different and that their personalities cannot be classified in everyday language. George Bernard Shaw once said "The reasonable man adapts himself to the world: the unreasonable one persists in trying to adapt the world to himself. Therefore, all progress depends on the unreasonable man." Perhaps his message was to tolerate the dissenters, the faddists, the kooks, and in general those who disagree with what we know is right—so long as they don't become too violent. Tolerate them, not out of any sense of humanity but for crass self-interest. A few of them are innovators, and society needs them infinitely more than they need society.

In reviewing all the groups of scientists and science-trained individuals we have encountered, we find a range of individuals spread over the whole spectrum of human behavior but with some important common characteristics. They range from individuals who could easily blend into a meeting of the Rotary Club to others who would find such a meeting more strange and confusing than listening to Little Red Riding Hood discuss comparative anatomy. Scientists are neither supermen nor naïve children. They are not foggily absent-minded or unrealistic; rather, many of the things they consider important and real are often quite different from those of the "everyday" world.

The Scientific Subculture

We have spoken of the scientific community as a subculture, and in a very real sense it is. A subculture can be identified as a group that differs in some significant ways from the larger culture and yet has enough in common with it to keep it from being totally distinct. It is perhaps easiest to distinguish subcultures by comparing their value systems and their rules governing behavior. Since many subcultures have a number of characteristics in common, identification can be made best in terms of a total profile rather than a single particular criterion.

There are a number of characteristics of modern science that have had a very decided influence on establishing and enforcing the prevailing value system within the scientific community. None of these are uniquely characteristic of the scientific community. The scientific subculture is characterized by: (1) a severe and formal selective process; (2) disciplines within the subculture involving small numbers of active participants; (3) far-reaching communication and mobility; (4) an internal system for establishing relative status; and (5) a set of formal and informal rules that govern status and support the value system. Although the fourth and fifth points are included in this list, they will be discussed separately later. Characteristics of status, values, and rules are closely intertwined, and their separate classification is largely a matter of convenience.

The severe and formal selective process that precedes one's acceptance as a scientist limits the number of possible participants in the scientific community. It also makes for acceptance of and adherence to the tribal mores. Scientists go through a period of five to nine years of indoctrination in their particular discipline. The portion of this time spent in graduate study is devoted to close contact with individuals who have undergone a similar indoctrination. In addition to the formal requirements, graduate programs usually force the student into close social contact with others who are in similar circumstances. His world comes to revolve around his particular discipline. There are few experiences, other than taking monastic orders or going to certain professional schools, that give this degree of orientation to a discipline. It would be surprising, indeed, if a person did not respond to the values and rules of a system that plays this large a role in his entire life. One of the obvious effects of such total submersion into the system is that the associated values assume a particularly large role in developing the individual's view of life. Formal and informal educational forces that mold the scientist will be discussed in the section called "Stumbling into Science."

A second characteristic of the scientific community is that the number of individuals working in a particular problem area is relatively small. For example, in the authors' separate specialties there are probably about 200 investigators in one and 100 in the other. If we consider only those who are directly interested in their particular research problems, the numbers may be 20 and 15, respectively. It is typical that these individuals are acquainted with the work of most others with similar interests and very often establish personal contact or correspondence. One of the principal functions of conventions and professional meetings is to maintain these personal contacts (and usually much more information is exchanged and more interesting ideas explored in the bars, in hotel rooms, and at parties than in the formal convention meetings). Another product of the personal relations within disciplines is that theoretical issues very often have strong personal implications. Again we emphasize that science is not as impersonal as often pictured. Attachments and disputes can take on intensely emotional tones and may be conducted with the dignity and logic of a fishwives' wrangle.

The third characteristic involves the wide-ranging communication and mobility of scientists. As mentioned before, science is not a completely solitary game, and scientists tend to flock together like geese during migration. Very often these concentrations ignore national boundaries, and the communication and contact lead to establishment and maintenance of an international system of values and rules. In the early 1900s, when Germany was the leading scientific center of the world, many scientists were attracted there to study and often to remain. From the end of World War II until about 1970 the United States, because of its lead in many areas (due in large measure to money and safety), attracted researchers from all over the world. Recent cuts in scientific funds have greatly curtailed this exchange. Certain institutions also attract many researchers. There are only a few institutions that are important in areas such as nuclear physics, astronomy, and oceanography. Even in disciplines where expensive installations are not important, the presence of one or two important theorists or researchers makes an institution very attractive to the young scientist. There are other aspects to the mobility of scientists. Within rather wide limits, scientists can move freely from one institution to another without seriously disrupting their work or being forced to re-establish their reputation. In contrast to most fields of endeavor, the scientist is not primarily dependent on local evaluation or evaluation by those unacquainted with his craft. Due to ease of communication and personal contact, his judges are his peers wherever they may be. It is a wise dean or department chairman who realizes that his own judgment of a scientist is of slight importance unless it coincides with that of the scientist's peers.

Values

Earlier we discussed some of the ways in which a scientist acquires his particular value system, without specifying what values he acquired. There are a number of highly specific customs peculiar to any specific discipline. For example, in physics it is generally accepted that observations are to be put in mathematical form at some point. On the other hand, such a rigid ground rule would seem strange or perhaps stupid to the anthropologist. In each case the scientist is acting in accord with the tribal customs, or "values," of his discipline. Fascinating as the variations among disciplines might be, we will consider only those values that cut across a wide variety of disciplines.

One of the values is individuality. The scientist is expected to reflect his own judgment and analysis within his field. This doesn't mean that his views must be different from those of others; they don't have to be unique in every respect, but they should be his own views, not mere echoes.

Another characteristic valued within many sciences is the ability to consider the data and interpretations without personal involvement. The principle is that the scientist should be able to back off and look at the data and interpretations as though they had come floating in a bottle from the sea. There are some scientists, personified by theorists, who definitely do not fit all parts of this description, but it remains a value even among them despite the violations.

One of the most important values is the dedication of the individual to his discipline. The system is directed toward fostering dedication; laziness or lack of interest meets with all the sympathy it would have received from the early Puritans.

A question that is often raised in relation to values is why more scientists do not have, and implement, ideals that would lead to resolution of some immediate practical problems. The problem is not one of ideals but of behavior. As indicated earlier, many scientists are interested in knowledge as an end in itself. In other words, they are trained and interested in problem solution, not solution-application. Many feel that solutions for problems such as war, poverty, and education already exist. The problems of solution application are seen as political and social. The critical difficulty is that scientists attacking an applied problem directly are required to move outside their paradigm. This is like asking an accountant to become a salesman, an economist a businessman, or an architect a brick mason— which can happen but rarely does and even more rarely is successful.

We could consider a number of other items in this section on values— for example, honesty in research is probably the single most important value in all sciences—but because of a high degree of overlap, we shall discuss these other items in the later sections on status and rules. Both status and rules reflect underlying values, and adherence to values is rewarded through achievement of status.

Status

The scientific community, as is true of most subcultures, has its own internal system for establishing status. Its status hierarchy derives from its value system; however, we are not primarily concerned here with the source of status hierarchy. Status in different communities can be based on a wide range of criteria. Among so-called "society," ancestry gives status within the group. Among some South African groups, obesity gives status. A skinny leader is absolutely unthinkable. Among businessmen, wealth (or the appearance of wealth) is a primary status symbol. Among the Plains Indians, the number of horses stolen was a source of distinction. There are even status hierarchies among other primate groups. For example, among mandrills, male status is related to the brilliant blue of their faces and the equally brilliant red of their behinds. The status of an individual may be due to a characteristic that is arbitrary (ancestry), achieved (horses stolen), accidental (tendency to obesity), utilitarian (wealth accumulated), or hereditary (mandrill's coloration).

In science the status hierarchy is a combination of a number of factors, including the following: (1) the discipline of the scientist: this is largely arbitrary but is related to the stage of development of the discipline; for example, physics rates higher than sociology; (2) the role played: a theoretician generally rates higher than other scientists; (3) originality and influence on others: influence on others ordinarily takes precedence; this order is reversed only when the original work has sufficient impact to influence others so that it becomes a combination of originality and influence; (4) the institution with which the scientist is associated: the hierarchy may be difficult to define, but most scientists can identify it. One measure that correlates with but does not define the hierarchy is the amount of research funds available to the institution.

Rules for Behavior

In addition to the positive aspects of achieving status, there are some informal prohibitive rules prescribing certain behaviors; these are derived from the value systems. Violation of these rules ordinarily results in various degrees of loss of status or rejection by the scientific community. These rules are not exclusive to the scientific community, but the great emphasis placed on them requires separate discussion.

The single most important rule concerns honesty within any discipline. A scientist can be an alcoholic, a braggart, a lecher, a thief, a murderer, a traitor (outside his discipline), or even a member of a civic club and still be forgiven, but one instance of his deliberately falsifying data brings eternal condemnation by his peers. The severity of this penalty together with the individual's indoctrination in the value system tends to ensure a high standard of professional honesty. An additional restraint is that science, being public in nature, allows checking of data by uncommitted peers. There have been some instances of violations of the rule of honesty; an example may indicate the attitudes involved. Several years ago a doctoral student published a dissertation which alleged some very important medical effects resulting from the use of certain chemicals. The report caused a sensation in certain circles. Other scientists repeated the study with negative results. After some time a representative of the national organization within this discipline investigated the original data. It was found that the data had been deliberately falsified; the student had not even visited some of the medical institutions where the data was reported to have been collected. An extensive report of the investigation was published in the discipline's scientific journal. At last report the former student was teaching in an obscure rural school. It is doubtful that any respectable journal would now even consider accepting an article from this individual. Destruction of an intelligent human is at best a sad affair, but the integrity of science is more important than the career of any single scientist.

The second most important prohibition for scientists is related to the Commandment "Thou shalt not steal." The important point for the scientist lies in the objects that might be stolen. There is no special scientific admonition about stealing bread or gold, but there are strong rules against stealing another man's work or ideas. In a recent case, a professor at a major university lost his position and was admonished in a scientific journal as a result of publishing another man's work without giving him credit. With the exception of elementary textbooks and those dealing in very general statements, it is a gross violation of scientific ethics to discuss another's work without giving specific credit. Even in these exceptional cases it is considered good form to give some important general references. There are good reasons for citing the work of others: (1) it helps show continuity in the development of ideas; (2) due credit is given to the originator of a particular item; and (3) most important, the serious reader can go to the original sources. Often a writer misunderstands or misinterprets the original source; citation leads to greater care and allows the reader to make his own interpretation.

Theft of ideas is a very thorny problem. In science, as in any creative endeavor, a man's ideas are much more important than mere money. The problem sometimes arises when discussions between colleagues leads to the emergence of an idea; often there is no possible way of establishing who said or developed what. Many of the present authors' discussions about the writing and editing of this book have gone on at a furious pace, with both often talking at once. Under such conditions we could each deny responsibility for errors and claim as our own any new contributions. The only obvious solutions are trial by combat or assumption of joint responsibility for all matters. As self-proclaimed and confirmed cowards, we choose the latter course. It is even more troublesome when, as has frequently happened, the same idea occurs independently to two scientists. Decisions about priority in such cases are almost impossible. You may recall the cases of Darwin and Wallace (evolution), Newton and Leibnitz (calculus), and DeVries and Mendel (genetics). Unfortunately, some cases have been handled with less than complete civility.

There are within science those who have few scruples about stealing ideas or have such convenient memories that they "become" the originators. These people are generally known, and only the naïve discuss anything more important than personal gossip, sex, sports, or the weather with them.

The Choice of a Research Area

The choice of a research area may seem an unlikely subject for a chapter on scientists as people, but please be patient. Research covers a large number of popular topics and a larger number of topics with only a few interested participants. Why do particular areas of research exist? Is it because the natural phenomena simply cry for investigation? This may be part of the answer, but at best it is only a part. Researchers often create an area by showing that a seemingly limited topic contains a variety of interesting phenomena. There are topics that have been abandoned, others that attract only one or two researchers, others that have yet to be opened, and still others where the researchers swarm like gnats. Why is a research area attractive? A particular area may appear to offer the possibility of posing and answering interesting questions. In addition, a particular topic may fascinate some individuals. A complete explanation for the attraction to different research areas would be impossible. The question is like asking why a man finds a particular woman attractive; the possible explanations are unlimited, but they are not likely to be conclusive.

The vast number of possible research areas that have received little or no attention may suffer for one of three reasons. First, we may not know that the subject matter exists. There is no cure for this lack of awareness other than tripping over it in our wandering, as Leeuwenhoek did in the discovery of bacteria. Second, we may know of the problem but lack the technical capability to investigate it. For example, scientists assumed that the moon had a backside, but prior to the development of rocketry the possibilities of direct observation didn't exist. Third, although we may know of a problem and have the techniques to investigate it, few become interested in it. Why? Who knows? Some problems are studied by only one or possibly a handful of investigators. How do we account for these hermits of science? If one is the only investigator in a field, the problem of competition does not arise; for some, safety is important. It is also possible that a Player develops a feeling of proprietorship in a particular area when he is the primary investigator. In such a case he may even resent the intrusion of other investigators and may discourage their entry.

Our problem is further complicated by fads and fashions. There are sudden fads in science as well as in the everyday world. For a time a particular topic may get "hot" and investigators swarm in. There are a number of investigators who are always entwined with the latest fashion. Many other investigators find a topic that suits them and ignore the fashion changes. A new topic generates concepts that may prove fruitful or may become a will-o-the-wisp. By the same token a steady, self-chosen course may only lead to a deeper rut. The game is not without a large element of chance.

Most scientists could have chosen any of a number of different disciplines. The election of a particular discipline from a fairly large set is likely to be fortuitous. Like marriage, proximity at the right moment, some undefined attraction, convenience, or the absence of competition can all be cited as reasons. In science we might find that a book, an article, a lecture, or simply the path of least resistance lies behind the decision to work in a particular area. Again parallel to marriage, there is a difference between the initial decision and a continuing relationship. The continuing relationship can be based on genuine delight in the relationship itself. There is also the possibility that the relationship is controlled by inertia, fear of new experiences, or indecision.

Stumbling into Science

This section is called "Stumbling into Science" because, like choosing a research area, becoming a scientist has a large element of chance. Despite this element, however, we can look at some identifiable characteristics of people who are likely to become successful scientists. There are exceptions to each of the statements that will be made, but these statements do seem to be true of a large group of successful scientists.

The prime requirements for the successful scientist are curiosity, intelligence, an interest in understanding concepts, and the capability and motivation to work long hours. Alas, such attributes, when found in the young, are often threatening to their teachers and classmates. In addition to the prime requirements, the prospective scientist is likely to be independent and somewhat aloof. These are not the general characteristics that lead to nomination as most popular student. Surprisingly, in spite of the apparent handicaps, prospective scientists tend to be rather well adjusted socially.

There are no formal rules for recognizing or measuring any of the above requirements except intelligence. This can be measured relatively reliably by scholastic aptitude or intelligence tests. There has recently been a controversy over how to interpret intelligence tests. However, for those who have no obvious cultural deprivation—and perhaps even for those who do—the tests do predict rather well if they are given and interpreted by competent testers. For most doctoral programs in science, an I.Q. of 120 or above is required.

Formal Education

Today the process of becoming a scientist is much more formally defined than in earlier times, although there are differences from institution to institution and among disciplines. Some of the general statements that follow do not fit individual cases, but they will give some idea of what is involved.

One or two centuries ago the scientist was often a gentleman-amateur, and even those who had formal training had received most or all of it as apprentices. The gentleman-amateur now either is extinct or has assumed the role of spectator, and apprenticeship alone has become a very difficult route to science.

Today, in order to have a successful career in science, it helps to start young. The most successful scientists are apt to learn a lot of math and read widely when young. They usually get good grades in school and go to top-rated colleges and universities. They show diligence and persistence throughout school and score high on standardized tests, particularly on the quantitative parts.

The formal aspect of science education ordinarily begins at the undergraduate level at a college or university. A future scientist needs a broad educational background. He should have a good orientation in such fields as the arts, history, philosophy (particularly twentieth-century philosophy), literature, and mathematics. He should develop a familiarity with computer science. If he has a knack for languages, they may help him; otherwise, he should avoid them. The two years of required language training as an undergraduate will be largely useless if he stops there. If, however, he is fortunate enough to get an extensive background in Russian, Japanese, or Chinese, he will find them very useful. German and French are often recommended, but they are not essential any more. Hitler put German science on about the same level as Bulgarian, and French science fell on woeful days long ago. There are disciplines in which these statements would be vigorously and profanely challenged. In any event, good translations of the literature from these two languages are generally available. Also, many people read these languages, and only a limited number read the others recommended.

A future scientist must get into a graduate school—the higher quality the better; this requires good grades in college. A *minimum* average of B is

required to be accepted in most graduate schools, and it is often necessary
to have some A's in solid courses such as math to accompany a B average. A
failing grade in any hard-core courses may be fatal, whatever one's average.

The undergraduate scientist should not attempt to become a specialist;
that will come soon enough. A scientist should be *educated*, not just
trained. Unfortunately, not all graduate schools admit students according
to this principle. The plea here is for a liberal education at the undergraduate level, not in the classic mold but attuned to the broad reaches of
current knowledge and thought. This sort of program should precede
specialization so that one has at least the capacity to recognize the existence
of different paradigms and the relations between various fields. The alternative to a liberal education is often like a typical business-administration
program, in which the student gets approximately twice as many specialized
hours as does the physics major. Perhaps business is more complex than
physics; or perhaps the object of a business school is to produce the
equivalent of a trained seal. Among their other excellent qualities trained
seals perform on command, present a uniform, neat appearance, and slide
smoothly through the water, creating no waves.

Probably the most important step in the formal education of a scientist
is his graduate program. He should shop as carefully as possible for an
institution and should try to get information from recent catalogs and
faculty members, since departments change very rapidly. He should pick a
department with a number of active young researchers. There are departments composed primarily of grand old names; if the history of science is
one's interest, this type of faculty has many advantages; otherwise it should
be avoided. By the end of his first year of graduate work, the student should
have a pretty fair idea of where his specific interests lie. He is also likely to
become the apprentice and junior colleague of some faculty member.
Hopefully the instructor will present an outlook that inspires the student
to make science a lifetime hobby as well as work.

Most of the factual material a student is given through undergraduate
and graduate education will rapidly pass out of date. A Ph.D. who drops
out of a field for about five years usually requires substantial retraining
before he can return to basic research. (Of course there are relatively static
areas where this statement can be challenged.) The moral should be clear:
The principal objective of an education is to prepare the student for

continuing self-education. Incidentally, the problem of self-education is closely tied to research. The active researcher is forced to keep reasonably current in his field. The researcher who is not familiar with what is going on in his field will have his omission pointed out by an editor or colleague. Editors, in particular, can be rather nasty.

Some Informal Education

There is at least one part of informal education that is important to the undergraduate, graduate student, and faculty member—the bull session. As you know, the discussion of a science, or any academic discipline, is not always the central topic in a bull session, particularly at the undergraduate level. Consideration of the more popular topics must be left to other books, since this one is about science. The potential scientist will find it helpful to spend a portion of his time in bull sessions. Some of these sessions will be enlightening, some confusing; some will give vital data regarding individuals, some will be drunken and muddled, and some—well, who knows? The real functions of these sessions, no matter what the topic, are those of testing ideas, informing participants of the lore of culture, trying fledgling wings in competition, and organizing individual thinking. All are quite reasonable functions and worthy of some time, so long as the student doesn't make a career of such sessions. As he progresses, the student will note that the subject matter tends to change somewhat. To be sure, sex continues to receive reverent attention, but other intellectually stimulating topics will begin to occupy a significant portion of the time. As the group changes and as the student and his peers become more oriented to a discipline, the ideas, confusion, and anxiety surrounding the discipline will increase in importance. At the graduate and professional level he will find that "shop talk" has become the principal focus of the sessions. Other ranking subjects will be politics, athletics, social changes, and the stupidity of certain segments of the population, both scientists and nonscientists. Scientists are usually criticized much more severely than other people. Humor rates quite high in these exchanges. Under no circumstances should one take himself or others too seriously; there is nothing worthwhile that is completely serious. Admittedly, some of the topics may resemble gossip more than formal theory, but this too is an integral part of one's involvement in the discipline. A young student is particularly fortunate if he has the chance to participate in a session with one of the legendary figures of his field. The results are usually illuminating, whether the giant turns out to be living in the distant past or the equally distant future. In short, this informal activity is an important part of education.

Some Final Thoughts on Education

The fortunate student will have an instructor who makes his subject live. It is very tempting to follow the lilting tune of this Pied Piper. And probably no harm will be done, since his interest is likely to be contagious.

For the student who has capabilities of becoming a scientist, many specific disciplines could provide a source of unending entertainment. If possible, the decision should be delayed for a time. Once he has arrived at a decision, the student should head straight for the laboratory and library. He should wash test tubes, clean cages, bring in coffee, build equipment, prepare specimens, sweep, mop, compute results, and read, read, read; in short, he should do any and all the jobs he can, whether paid or unpaid. This may not be the perfect test of his interest in the particular discipline, but it will give him first-hand knowledge. A caution: many universities and researchers do not encourage undergraduates to get into the research laboratory; the student should be patient and willing; free help will not go unrewarded forever.

Reading cannot be overemphasized. The student should not limit himself to his chosen area. A wide range of reading will not make him an expert in every field, and it may not even be any direct help in his chosen field. But it is a good safeguard against becoming simply a well-trained technician.

Employment of Scientists

The previous section, about how and why people go into science, was based on our own view of the field. How scientists are employed can be tied more closely to data. When this section was written for the first edition, nearly all groups and individuals (the authors included) felt the problem was to fill the demand for scientists. Today the problem seems to be more one of finding the demand. The entire situation is quite complex and reams have been written bewailing, denying, and explaining. We will try to hit a few high spots and hope interested readers will look into it further (perhaps by reading Cartter's article).

One of the most troublesome and difficult aspects encountered in attempting to make plans for future scientific needs is that both institutions and individuals must always project several years in advance of the actual events. For example, the average time between entering college and completing a Ph.D. degree is nine and one-half years. The student has some flexibility over the first few years, but as time passes, his options become more limited. The lag time for institutions is also very long, which affects starting, stopping, or changing the direction of programs. The fact that some fields, such as nuclear physics, changed from a shortage to an oversupply in a period of about three years is enough to make all involved wary. A number of other factors contribute to making our little guessing game resemble blindman's buff. Future economic conditions, governmental attitudes and commitments, student preferences, and public attitudes, in addition to changes within or affecting science directly (new discoveries, new fields, new technologies), all complicate the situation.

In spite of these difficulties, the situation must be dealt with. A few suggestions come to mind immediately. We do need some up-to-date and reliable figures as to just where we stand. Many institutions are in extreme financial straits. Cancelling expensive Ph.D. (or other) programs may be both rational and timely. Governmental agencies, particularly the National Science Foundation (NSF), need longer-range planning authority. The list could go on but since the main objective is to outline the problem, perhaps this is sufficient.

Two other items are noteworthy. Scientists do still get jobs—not always what they want and sometimes outside their specialties, but actual unemployment is lower than among the remainder of the population. While some fields have been particularly hard hit, others have been only slightly affected. Second, there is a serious danger that failure to continue recruiting young scientists in the academic area may decrease the probability of innovation. No matter how loaded with honors an elder statesman of science may be, it is still the brash young upstart who is most likely to produce something new.

Should the foregoing discourage the prospective scientist? Not necessarily. Competition will be tougher, and the marginal student will find both graduate school and the job market very difficult. The dedicated (or driven) student of high capability should continue to find satisfactory chances to become a part of the ongoing enterprise. (NSF's 1970 *Review of Data on Science Resources* indicates that the percentage of graduate students enrolled in nonscience areas is increasing faster than in science areas, and that this is a long-term trend that began about 1963. The meaning of the trend is not clear.)

Most of the following data are taken from the *National Register of Scientific and Technical Personnel*, 1970. Copies can be obtained from the Superintendent of Documents, U.S. Government Printing Office, Washington, D.C. 20402.

As you will note later, the definition of "scientist" used in this section differs markedly from the one we used previously, because the paradigm here is that of the United States government. The text points out the differences where definitions are involved. Most of the 1970 survey was based on questionnaires submitted to scientists. This also creates problems. Despite the glowing description of scientists you've been reading here, they too have their vanities. You may recall the statement that basic science is higher in the status hierarchy than applied science. If you, as a scientist, work in both applied and basic science, are you more likely to describe your "primary work activity" as basic or applied? Two guesses. Further, the salaries mentioned vary greatly from institution to institution, by geographical areas, and by specialties within disciplines. Here we will ignore such important considerations in favor of general tendencies.

The 1970 survey covered approximately 313,000 "scientists"; of these, about 40% held Doctoral degrees, 27% Master's degrees, and 30% Bachelor's degrees. Whether those with degrees below the Master's level should be considered scientists or technicians is an open question.

Of the group surveyed in 1970, 42% were employed in educational institutions ranging from junior colleges to universities, 10% were employed in various governmental entities including the military (about 2%), and 31% were involved in business and industry. Other sources of employment (13%) were too scattered to fall into any one category. The rate of unemployment is estimated at about 4%, a substantial increase over recent years. A great part of this increase is due to changes in government funding.

There were more chemists (8%) than any other single category of scientist in the survey. Anthropologists, linguists, sociologists, and statisticians each constituted 1% or less of the sample. One interesting aspect of the survey is that apparently only about 7,300 MDs were classified as scientists. This number represents about 2% of all MDs. Considering the scope and importance of medical areas, this representation is pitifully small. On the other hand, when you consider that these 7,300 have voluntarily accepted an income of $10,000 to $15,000 per year below that of their classmates, some sort of tribute seems in order.

While we are on the subject of money, let's briefly consider scientists' incomes. They rarely get rich, but on the whole they are reasonably comfortable. Again the situation is complex, and we can look at only a selective summary of the data. In academic institutions the survey indicates that the median salary was about $13,500 per academic year over all areas of science. The range is not great between the lowest-paid group (mathematics, $11,900) and the highest-paid group (anthropology, $14,000). As might be expected, pay was better in industry (overall median, $16,700) and better yet for the self-employed (overall median, $20,000). The difference between academic pay and industry was most certainly understated, since the scientist entering industry is less likely to have an advanced degree than the academician. For example, a substantial number of individuals in the survey had a Bachelor's degree; but a Bachelor's degree is rarely acceptable for academic ranks. In universities, a very large proportion of the science faculty have Ph.D.s; at the university or college level, a Master's degree is almost always a necessity. The civilian portion of the federal government was paid about the same (overall median, $13,000) as those in educational institutions. The most lucrative self-employed fields were statistics and psychology; the median for these disciplines was approximately $25,000.

Activities of Scientists

The NSF surveys provide data for a rough estimate of the primary work activities of scientists. The major activities are research and development (R & D, 31%), management or administration (22%), and teaching (23%). The R & D item needs somewhat closer examination. The best estimates we can make are that less than 10% of all R & D activity is devoted to basic research, while the remainder of those in R & D work in applied research or development. These estimates represent a compromise between the NSF manpower survey and their *Review of Data on Science Resources*.

Some mention should be made of the scientist-turned-administrator. Combining active research with administration is a rare and difficult situation. The problem of administration in science is quite serious. Administrators are not likely to be among the current leaders in research, and only the current leaders are likely to be capable of fully understanding the research problems. One solution used by some effective administrators is to use active researchers as advisors.

The Daily Grind

Research scientists as people are much like the rest of us. They eat, cannot find a pair of socks or stockings, curse the idiots in traffic, exchange meaningless polite greetings, admire a miniskirt, change a diaper, remark on the undeniable cretinism of bureaucracy, attack sexism, and in general behave like tame creatures from our society. Then there are things characteristic of their occupation. They think about problems, design experiments, talk to research assistants and technicians, look at equipment, check apparatus, cope with malfunctioning equipment, analyze data, puzzle over confusing or conflicting data, discuss problems with colleagues and assistants, hold meetings or seminars, worry about research budgets, and read, think, and write. These last three activities are their most important functions; if scientists are fortunate, these activities occupy a substantial portion of their time. If they are academic scientists, somewhere in the schedule they meet their classes and try to communicate some of their knowledge and enthusiasm to their students. If we are all fortunate, they may contribute something to the acquisition and dissemination of scientific knowledge.

Of course, not all is perfect. Science, like any other profession, requires a great deal of plain hard work. The ten-hour work day is more common than the eight-hour day. And not every day is glamorous or exciting. Sad to report, frustration is a more common experience than exhilaration. The scientist may become bored, confused, tired, fed-up, and angry. He may conclude that the whole game is pointless. Fortunately, these states are usually transitory; on the balance the attractions outweigh the blemishes.

Chapter 7
Science
in the World

Numerous discussions in this book have indicated that science has a very strong impact on the everyday life of nonscientists as well as scientists. The effects are seen in the way people view themselves and their universe (the intellectual aspect) as well as in more obvious technological progress. Even the attitudes and views of the scientifically naïve are influenced by science, although these people are generally unaware of the source of this influence. Galileo gives us an excellent example of both the intellectual and technological influence of science even in his day. Galileo's astronomy was of the utmost importance in changing man's view of his place in the universe and in the struggle between science and authority as bases of knowledge. On the technical side, Galileo's improvement of the telescope and studies of ballistics had an undeniable practical impact.

Science and Social Change

Over the past 100 years it has become increasingly apparent that man has the capability to design or change both his material and his social environments. With the rapid development of science and technology, the number and range of possible alternative decisions he must consider has increased fantastically. The capability for change is still not absolute, nor will it necessarily lead to a better system. (Read Huxley's *Brave New World* and Orwell's *1984*.) One important impact of the change in attitudes wrought by science is that, except for the political Neanderthals, no one still believes that any particular system is "given" or is the only possible one. Man now has to face the painful fact that he can and must make decisions regarding his material and social environment, and often he must do so without any precedent for the decision. A decision is always made, since failure to act is also a decision.

Scientific Advancement

Let's look at an example of a probable future development to see how this development would lead to problems and changes in the intellectual, social, and technological aspects of our society.

Take the subject of aging. Let's assume that aging can be controlled so that an individual can have a useful life span of about 200 years. Incidentally, such a development seems likely, provided the major powers forego the ecstasy of mutual extermination. Let's also assume that this increase in life expectancy could be accomplished within the next fifty years. (We will ignore the logical problem of how we would know we had extended life by 100+ years before that interval had passed.) It is estimated that the total scientific and technological development necessary to achieve this goal would not cost more than a yearly expenditure equal to 5% of our present yearly military budget. Recognizing the relative attractiveness of life destruction versus life preservation, as well as the immediate self-interest of some groups, we can't expect any great diversion of funds. Even if military expenditures are decreased, a cry of "we can't afford it" will probably block or delay any large-scale effort. But ignoring this pessimism (or realism) concerning funds, assume that at some date we will control aging. What would be some of the intellectual and technological results in our society?

From a technological viewpoint, consider some of the economic consequences. Any impact on economic theory per se will be ignored here, since our purpose is one of illustration. One very obvious economic consequence of controlling aging is that of training people in the work force. At present the typical individual completes most or all of his formal education prior to full-time entry into the work force. But even today it is necessary to send engineers and other technical personnel back for refresher courses on some occasions. New developments must be communicated or else technical personnel rapidly become outmoded. The same thing is true to an even larger extent among scientists. For example, the social sciences recently developed use of computers, statistical techniques, and mathematical models, leaving many older scientists bewildered, since they don't have the background to understand these new developments.

Consider the plight of someone who completed his education about 1850. Assuming he were alive today, could he be qualified to perform any complex task in today's economy through self-education? Undoubtedly, a few individuals could; in most cases they could not. Such problems seem certain to become much more evident and more serious when aging is controlled. New concepts of education would be necessary. Instead of an early period of education followed by a few "refresher" periods, a regular routine of return to education may be required. We could follow the educational problem further, particularly the question of how to keep the educators themselves current, but even this brief a discussion should suggest a host of related problems.

It has been said of politicians that "few die and none resign." Industry has been somewhat more successful in maintaining a turnover in top management, primarily through the use of arbitrary retirement ages. One important part of our present problem is that long-term incumbents (political, industrial, or otherwise) tend to use old solutions for new problems. In addition, long occupancy of an office establishes a feeling of ownership together with the power to defend that ownership. Is it desirable to have the president of a corporation or university remain in office for 100 years? We think not. The effect of very extended periods of tenure in advanced positions would also be felt in the ranks of young aspiring workers. How many of these would have the patience to wait 20, 30, 40, or 50 years for promotion?

The control of aging will surely require some basic changes in our economic organization. It would serve little purpose, other than the joy of speculation, to consider a wide variety of economic problems accompanying the control of aging. The examples given should indicate the breadth and complexity of these problems.

Our present attitudes toward something as basic as human life are related to the period of growth and to life expectancy. Although it seems certain

that many of these attitudes would change, it is not clear exactly what form
such changes might take if aging were controlled. We can look at a few
problems, however, to see the sort of attitudes that would be influenced by
the control of aging.

Since men ordinarily don't live more than about 70 useful years, extension of this period to about 200 years would have to be controlled through
some sort of specialized treatment. The first obvious question is, should
everyone get the treatment? Or more bluntly, who lives and who dies? The
first impulse is to answer that everyone has the right to live. But what of
the individual who has suffered permanent and serious damage, deformity,
or other disability through illness or accident? Should he be preserved if
there is no serious hope of recovery? What if he is in intense pain? Or has
become a human vegetable? These may seem like extreme cases, but they
do illustrate some of the difficulties. Other cases that arouse mixed feelings
are individuals who pose a clear threat to society. Do we prolong the life of
a person imprisoned for a violent crime? The result might be a choice between an endless sentence straight out of Dante or the guarantee of release
at some point without regard to the consequences for others. These problems exist today but would become much more serious if life were extended.

In the early stages, and perhaps for a long period during a program of age
control, it is quite unlikely that facilities or personnel would be adequate to
extend the program to all the people in the world. Would life itself become
a new tool in nationalistic competition? If facilities were not sufficient to
take care of entire populations, how would decisions be made? How do you
decide which life is worth preserving? A few years ago a somewhat similar
problem faced a group charged with deciding who would be scheduled to
use a limited number of artificial kidneys. The choice was literally one
between life and death. The agony involved in these decisions was obvious.
Multiply this by millions and consider the impact.

Admittedly, the first examples have been chosen to indicate some sharply
defined effects of our hypothetical development. Other effects might not
be so dramatic, but they could conceivably have as great a total influence
on our attitudes. How would men value life if they had the prospect of 150
or more years of useful activity? Would they cling to life with greater
tenacity, or would boredom dilute its savor?

Inevitably, population control would also become a problem even more
pressing than it is today. How would we ration procreation? Would it be

left to individual judgment? Would our attitudes toward new additions to the population remain unchanged? It doesn't seem likely. These questions don't have simple answers.

This speculation about age control has considered only a few facets of technological, social, and intellectual changes as examples. The possibilities for other changes are almost limitless. It seems highly probable that our entire society would be permanently altered, for better or worse. We must face the fact that, with or without major events such as the control of aging, our society is rapidly and constantly changing; the rate of change is accelerating and a substantial portion of these changes are due to science or science-related activities. Unfortunately, despite rapid change, our society suffers from lack of application of our knowledge and capacities, particularly in the social sciences. This failure to apply our knowledge is the principal topic of the next section.

Cultural Lag

As an example of the lag or failure to apply our knowledge derived from the social sciences, consider the problem of racial relations. In the early 1940s a Swedish institutional economist, Gunnar Myrdal, examined the situation of the Negro in the United States (*An American Dilemma*). In painstaking detail he and his associates observed the historical development and current status of the Negro in relation to economic status, education, social arrangements, political power, their attitudes, and white attitudes toward them in both North and South. Myrdal also drew conclusions on probable future development based on these observations. While Myrdal was obviously mistaken in some conclusions and possibly in some of his observations, the greatest part of his work has stood the test of time extremely well. A few quotations will give you an idea of what Myrdal described and foresaw. These quotations read almost like news stories today.

> The very presence of the Negro in America . . . represent[s] to the ordinary white man in the North as well as in the South an anomaly in the very structure of American Society. To many this takes on the proportion of a menace—biological, economic, social, cultural and, at times, political. This anxiety may be mingled with a feeling of

individual and collective guilt. A few see the problem as a
challenge to statesmanship. To all it is a trouble.

. . . the President made a solemn proclamation against
discrimination in the defense industries and government
agencies and appointed a committee, having both Negro
and white members. . . . Other branches of the Administration made declarations and issued orders against
discrimination. . . . The national labor unions also lined
up for nondiscrimination. The Negroes heard and read
the kindly promises. They again noted the public acceptance of their own reading of the Constitution and the
American Creed. But they knew the grim reality.

We are now in a deeply unbalanced world situation.
Many human relations will be readjusted in the present
world revolution, and among them race relations are
bound to change considerably. As always in a revolutionary
situation . . . there is, on the one hand, an opportunity
to direct the changes into organized reforms and, on the
other hand, a corresponding risk involved in letting the
changes remain uncontrolled and lead into disorganiza-
tion. To do nothing is to accept defeat.

. . . the equally troubled view of a Negro clergyman,
Dr. J. S. Nathaniel Troso: "I am afraid for my people.
They have grown restless. They are not happy. They no
longer laugh. There is a new policy among them—something strange, perhaps terrible."

We might note in passing that the paradigm of Myrdal and his co-work-
ers was quite different from that of the physical sciences; nevertheless, his
approach was that of a scientist. His collected data were based on intersubjectively testable observations, and he attempted to organize and interpret the data in some coherent way. Myrdal's conclusions don't have the
precision of Kepler's three laws of planetary motion, nor does he propose a
general theoretical organization; but he does analyze and attempt to explain some of the data he collected, and he used his data and concepts to
make predictions. His work meets the criteria for scientific investigation.

Many other examples of lags between the accumulation of knowledge
and its acceptance and implementation could be cited. A few will be mentioned briefly. Air and water pollution: we don't know all the answers, but
we can make substantial improvements now. Recent reports suggest that
although the quality of the air may be improving slightly, water pollution
is getting worse. Health: Between 1950 and 1962 the United States dropped
from sixth to eleventh place among countries of the world in the rate of

survival of newborn infants. More recent reports state that the U.S. has
dropped to fifteenth place. Several of our largest cities, including New York,
Chicago, Philadelphia, Detroit, Baltimore, Cleveland, Washington, and
St. Louis, have recorded increases of from 1.6% to 26.4% in infant mor-
tality over these years. Have we forgotten something, or do other countries
such as Ireland have superior medical knowledge? Let's not belabor other
obvious cultural curiosities, such as our refusal to accept the metric system,
our response to criminal or other antisocial behavior, our antiquated educational systems, our extermination of wildlife, and our resistance to fluorida-
tion of water.

But *why* are there serious delays between the accumulation of scientific
knowledge and its acceptance or implementation? Some of the general
sources of resistance to the acceptance and use of scientific findings can be
listed:

1. Difficulty of communication.
2. Inertia.
3. Confusion due to disagreement or apparent disagreement among
   scientists.
4. Influence of special-interest groups.
5. Fear of change.

First, there is the complex problem of communication. How many
people are likely to read, understand, or even know of a study such as
Myrdal's? A thorough knowledge of the scientific controversies over the
creation of the universe (big bang versus the steady state) is even rarer and
in addition requires an understanding of astronomy and physics that only a
handful possess. Because of such problems, the acceptance of scientific
statements becomes a matter of faith for most and usually requires a
generation or two to be effective, since faith changes slowly. Today most
people accept on faith the idea of the earth traveling around the sun, but
they are no more able to defend such a belief than their ancestors were able
to defend the belief that the earth was the center of the universe.

There are at least two other aspects of the communication problem.
Today there is a very serious lag in publication of new material. Often
"new" research is two or three years old before it is printed in journals. The
newest textbook material is generally three to five years old. Also, communication between basic and applied scientists is difficult, since these groups

read different journals, attend different meetings, and have different interests and backgrounds. Even when they attempt to communicate with each other they often fail, since they see the same phenomena through different paradigms.

Second, there is a problem of inertia. This is so common to any enterprise that lengthy explanation seems pointless. We need merely keep in mind that any change in the pattern of activity, thought processes, or an institutional organization requires extra effort. The direction of a rut may be irresistible.

Third, new knowledge may be surrounded by an aura of controversy. The disagreements may be quite basic, as illustrated by Agassiz's attack on Darwin's theory of evolution. On the other hand, the areas of disagreement may mask more general agreement on important issues. For example in the field of biological taxonomy, there are the "splitters" and the "lumpers." The splitters tend to divide and subdivide phyla, genera, species, and other categories into a large number of units and subunits. The lumpers use more limited numbers of divisions. Exchanges between these groups are often rather heated. Yet they generally agree on the kinds of data that should be collected, the animal groups that are related, the importance of tracing evolutionary changes, the mechanisms for these changes, genetic theory, the general principles of classification, and most other things. Application is often delayed when there is debate about the underlying principles, whether the debate is relevant to the application or not.

Fourth, all of us have the capacity to find a "good" reason for our resistance to change. These reasons often reflect only a special interest. It would have been surprising indeed had the tobacco industry agreed with the Surgeon General's report that smoking is a danger to health. Equally unsurprising has been Southern politicians' resistance to the Supreme Court's school-integration decision of 1954, because, in their words, it was based on sociology and not law. In these cases and many others, the special interest is rather clear but in some cases the special interest is less obvious. Economic, social, religious, educational, and other groups and individuals also have vested interests in particular aspects of the current scene. One would have to be a super Pollyanna or simpleminded version of Candide's philosopher friend Pangloss to expect that individuals or groups would readily give up their pet ideas, regardless of the contrary evidence or reasons. Whether people can learn to be more flexible or rational remains to be seen.

Failure or delay in coping with change will become an increasingly serious problem in a rapidly changing world. Flexibility and the capacity to accept rational change may very well be the margin between progress and disaster. This is not the place to resolve the problem. It is simply suggested as a lifetime research project for a few psychologists and sociologists.

A fifth obstacle that delays acceptance or utilization of scientific findings is the fear of change or new situations. There is very clear evidence that unknown or changed situations evoke both curiosity and fear. For present purposes we will ignore curiosity. The evidence of fear covers a wide range of animals, from rats to men. The fear produced by a new or changed situation may be as clear as that of a primitive man faced with his first airplane ride, or it may be the less defined fear that questioning or even discussing a favorite belief in terms other than clichés will bring down an entire belief system. It is difficult to define clearly the formless fears of unknown things. Perhaps a reasonable analogy is a child's fear of the dark. The fear may be expressed as "something might get me," or "there is a tiger in the room." Whatever the stated reason, it probably bears no relation to past experience. Fear of change seems to have the same irrational but real character. No matter how we state our objections to a new development, we should remain aware that part of our response is probably based on a very primitive form of fear.

While considering some of the obstacles to change, let's not assume that change for change's sake is necessarily good; rather, we must be aware that irrational resistance to change can prove fatal to an industry, government, economy, or any institution—even science.

Scientists are not a race apart. They also have built-in resistances to change. It has been said that scientists were not converted to the theory of evolution; rather, those who didn't believe in it finally died out. Although scientists do have some general rules for resolving arguments, change still generally depends on the younger members of the community.

The Relation between Science and Technology

The relation between science and technology is mutually supportive. Science benefits from technological advances, and much modern technology rests on scientific discovery. Until relatively recently, much of technology developed independently of scientific advances, although much of science depended on the then current technology. Today, however, new technology usually depends on scientific discovery, and this dependency will undoubtedly increase. In the past, technology was relatively simple and utilized readily available materials. This is unlikely to be the case in the future.

To illustrate the mutual relationship between science and technology, we might consider the microscope. An early step in the developmental sequence was the invention or discovery of a method of making glass. This event was preceded by the ability to control fire, as well as other necessary antecedents. Early records of the Egyptians indicate that they knew how to make glass, and this knowledge probably came from Asia. But the origins of this major technological development are buried in pre-history.

There is a report that a convex lens made of rock crystal was found among the ruins of the palace of Nimrod (about 1800 B.C.). Certainly, for many centuries prior to scientific research, men had known of the magnifying powers of natural or accidental lenses. A significant step in this history is a book of optics attributed to Ptolemy. Ptolemy did make significant contributions to science. His book on optics (written in the second century A.D.) laid the scientific groundwork for later work (about 1000 A.D.) by Ibn-al-Haytham, who advanced the science of optics through extensive research. Roger Bacon (1270+) was familiar with Ibn-al-Haytham's work and described a telescope, although there is no record that Bacon built one. The first modern convex lens (a simple magnifying glass) was made by grinding (a technological advance) sometime in the late 1200s. A major breakthrough in microscopes occurred about 1590 when Hans and Zacharias Janssen, Dutch spectacle makers, made the first compound microscope. In 1611 Kepler outlined the construction of an improved compound microscope, based on optical theory, and a model was built in 1628 by Christoph Scheiner. This microscope laid the rudimentary pattern for the modern optical microscope. Note that up to this time there was an intermingling of scientific and technological steps: the techniques of glass and lens making were necessary to the science of optics, and the scientific principles of optics were equally important to the design of an advanced microscope.

We could move from the development of the microscope to a discussion of its use, but it is sufficient to say that the microscope plays a very significant role as a tool of both science and technology. There are various discrepancies in treatments of the history of lenses and microscopes, but these discrepancies shouldn't disturb us since we are interested in the principle rather than the accuracy of this historical illustration. And the principle is not affected by minor variations in the sequence of events. If you find this acceptance of confusion about history strange, read H. L. Mencken's "Hymn to the Truth," which casts a beam of revealing darkness on the history of the bathtub.

To gain some appreciation of the technological products that rest on a scientific base, one has only to look around. For example, whatever the limitations of fluorescent light, it is vastly superior to kerosene lamps. Consider a few of the basic scientific discoveries necessary to produce this illumination. Again, the rude beginnings go back probably beyond recorded history. There is evidence that about 600 b.c. Thales, a Greek philosopher and scientist, rubbed amber with a cloth and observed that the amber then attracted lightweight objects such as strands of feathers. Doubtless others, in idle moments, had seen the attraction of amber for other objects, but we have no record of their observations; possibly they only shrugged and went on to more important matters.

The curious behavior of amber was known for about 2200 years before William Gilbert (an English physician) made an important extension to our knowledge of this peculiar phenomenon. Gilbert found that other substances (such as sulphur and glass) had the same properties as amber. Strangely enough, lodestone seemed to have some of the same characteristics as amber except that lodestone and amber attracted different bodies. It became clear that scientists could not limit their studies to the attractive qualities of amber. By 1646 Sir Thomas Browne (another English physician) contributed the name "electricity" to the phenomena being observed. (Browne also believed in witches and assisted in their examination, but he rejected lycanthropy.) Charles Du Fay (a chemist who was also superintendent of gardens to the King of France) studied the attraction and repulsion of charged objects; by 1733 he came up with the idea that there were two kinds of electricity. At this time (around the 1740s), as Kuhn phrases it, "There were almost as many views about the nature of electricity as there were important experimenters." The first and possibly greatest American scientist, Benjamin Franklin, proposed a concept that organized a large portion of the known data and gave direction to much later research. Franklin concluded that there was only one "electric fluid."

This fluid was found in all bodies; those bodies having an excess quantity were positively charged, while those with smaller quantities were negatively charged. The concept was crude and did not account for all the information available at the time, but despite its deficiencies Franklin's way of organizing the data was accepted as being superior to others. Franklin's successors developed his ideas about electrical phenomena in a number of ways, but his concept was a critical beginning. Franklin has been best known popularly for his later experiments and explanation of the similarities between static electricity and lightning (particularly the kite experiment, 1752). As ingenious as this later work was, it was not nearly so important as his earlier conceptualization. As pointed out before, concepts (or theories) are more important than isolated facts.

In passing we should glance at the work of the Italians Luigi Galvani (a professor of anatomy) and Alessandro Volta (physics), if for no other reason than the fact that their work originated in the twitching of frogs' legs. The anatomist concluded that frogs' legs contained electricity that was released when they touched metal. The physicist concluded and demonstrated that the chemical action of moisture and different metals such as iron and copper produced electricity. Note that the different analyses of the events is related to the difference in paradigms of the investigators.

In 1820 Hans Christian Oersted (a Danish scientist) showed that an electric current has a magnetic effect. By 1822 André Marie Ampère (a French physicist) worked out the laws on which the present-day concept of electricity is based.

Michael Faraday, the English chemist and physicist, gives a fitting climax to our consideration of the history of electricity. Faraday believed that if electricity could produce magnetism, then magnetism could probably produce electricity. In 1831 he found that relative movement of a magnet and wire loop leads to induction of an electric current. A lesser known American scientist, Joseph Henry, also discovered this principle in 1831. The principle discovered by Faraday and Henry is the basis of the construction of electric generators and motors. Why is it that this scientific history omits a discussion of Edison? He can be placed high among the developers of the uses of electricity, but his scientific contributions to the concepts of electricity are few.

Before drawing any conclusions about the scientific work discussed, let's

consider for a moment the technological fallout. Could a modern technological society exist without electricity? It's difficult to see how. Certainly
we could have developed alternatives for some things such as fluorescent
lights, but alternatives to radios, telephones, computers, T.V., radar, most
internal-combustion engines, and many chemical processes would be out of
the question.

From this abbreviated history of electricity we can extract a few points,
some of which have probably already occurred to you.

1. Investigation of electricity was truly an international and interdisci-
plinary endeavor. Greeks, Italians, French, Americans, English, Germans,
Danes, and Dutch all made significant contributions, as did philosophers,
amateur scientists, chemists, anatomists, physicists, physicians, and general-
ists. As has been mentioned before, science is an open community, with
membership dependent on contributions, not on politics, geography, or
professional titles.

2. The pace of the history picks up over time. The history is long, but
events come closer and closer together as we approach the modern era. This
increasing intensity is in part due to the interdependence of science. Each
new discovery or concept suggests additional possibilities or relates prior
findings.

3. Curiosity motivated most of the researchers. What distant horizon
hid the "practical" applications? Could Thales, Gilbert, Franklin, Galvani,
Oersted, or even Faraday have had immediate "practical" applications as
their goals? This motivation does not seem even remotely possible. These
men had found their game. Our present-day technological fallout is an
unsought and unforeseen by-product of their efforts. If their efforts had
been directed toward immediate application, where would we be? Not only
would we have failed to produce basic scientific knowledge of electricity,
but we could not have achieved our present technological development.
Had the Roman engineer-administrator outlook dominated our culture, we
would doubtless have achieved infinite refinements of existing knowledge
and techniques, together with some fortuitous gains. Just how we could
have achieved new insights and changes based on abstract principles is
impossible to conceive.

4. To restate an earlier point, technology and science have strong mutual
influences, even though they have quite different objectives.

Jerome B. Wiesner, Special Assistant for Science and Technology to President Kennedy, displayed courage, wit, and keen insight in his struggle to explain the nature of basic research and its possible relation to technology. Had you been in his position, using the example of electricity, how would you explain to budget administrators or Congressmen your reasons for financial support of Gilbert's amber-rubbing, Galvani's frog-leg preparation, or Franklin's kite flying? One's imagination fails at this point. Fortunately Wiesner's did not. His summary of the problem rings with the clarity of a rock-crystal goblet.

> The research scientist is primarily motivated by an urge to explore and understand, but society supports fundamental research primarily because experience has demonstrated how essential such work is for continued progress in technology. Halt the flow of new research and the possible scope of technical developments will soon be limited and ultimately reduced to nothing. Incidentally, scientific knowledge need not be exploited immediately once it becomes available. It exists for all to use forever.
>
> Thus, acquiring scientific knowledge is a form of capital investment. Unlike most other capital investments, it does not become obsolete nor can it be used up. Technological developments are also a form of capital investment, though somewhat less enduring. To be sure, a more efficient process will also yield its benefits endlessly, but usually technological developments tend to become obsolete as better methods, devices, and processes emerge.

Financial Support of Science

We now come to the grubby but vital topic of money. First, we must remember that science, on any significant scale, exists only in economies where there is a surplus of goods and services. A prosperous economy forms a necessary base for scientific work. Tragically, the economies of the "underdeveloped" countries, where the need is overwhelming, have not supported—nor have they been capable of supporting—any substantial scientific and technical training or research. Read the speech of P. M. S. Blackett, President of the Royal Society (cited in the Bibliography), in which the existing great disparity between the "have" and "have-not" economies is considered. The primary theme of his address is that the gap is not closing; on the contrary, it steadily grows wider.

The United States, as the most affluent industrial nation, spends a greater amount on research and development than any other nation. It is beyond the scope of this book to consider in detail whether the amount spent is adequate, either in relation to our own economy or in relation to our position in the world. At best the question is an open one.

In the last few years, financial support for science has changed significantly. It is frustrating to report that, in spite of numerous statements about the extent and nature of these changes, it has been impossible to find sufficient agreement to be sure just what has happened. Two points do appear clear. First there has been an overall cut in research funds, further reinforced by substantial inflation of prices. Second, there has been increased emphasis on "relevant" research—that is, research with immediate practical applications.

One report that reflects most closely our own estimate of the situation was made by a five-member committee from the National Academy of Sciences. This group investigated funding of chemistry projects by the National Institutes of Health. The committee concluded that funding had declined by an average of 20% and that the section most closely related to basic research had funds cut in half. The committee found that various institutes, in-house research, administrative budgets, and continuing extramural grants had been protected with the result that new and renewal grants had become a disaster area. Other reports stating that federal institutes' funds have had only limited cuts have apparently ignored the point made by the National Academy Committee.

What happens next? Some defend the cuts as "just trimming the fat," and others predict that either the Soviet Union or Japan will be the next great center of science. The political and economic climate of the next few years should give us some answers.

General Distribution of Funds

The following description of the support of research and associated activities in the United States relies primarily on the *Reviews of Data on Science Resources* issued by the National Science Foundation. The figures don't represent any single year, since it was necessary to quote figures from

different years for various reasons. But they do present a general picture
which, although dated, will give you some understanding of the situation.

Most reports published in popular media lump research and development funds together. This practice seriously distorts the picture and mis-
leads the public regarding the research effort. The National Science Foundation (NSF) uses three categories: basic research, applied research, and
development. Consideration of a few representative figures will show the
differences in these two approaches. When total research and development
expenditures were approximately $19 billion, only about $6 billion went
into basic and applied research combined. Applied research accounted for
about $4 billion, while basic research received about $2 billion. An analysis
of several years' budgets shows that distribution of funds was approximately
66% to development, 24% to applied research, and only 10% to basic
research. Development is primarily concerned with the technological applications of sciences. In this sense it is a gross distortion to combine the
research and development budgets, as is usually done in the popular media.

The distinctions between basic and applied science were pointed out in
Chapter 1. Our primary interest as scientists is the 10% of the total
expenditures concerned with basic research. These funds are the lifeblood
of science in the modern world. Incidentally, both authors are convinced
that the 10% figure for basic research is inflated and that approximately 5
to 6% is more accurate. The total amount devoted to basic research is quite
large when compared to earlier times, but it is still underfinanced. It is only
fair to state that the level at which basic research should be supported is a
matter of judgment; also, the authors' personal interest in basic research is
clear.

There are at least three points to be considered regarding the adequacy of
our present effort in basic research. First, the absolute amounts are de-
ceptive. Present budgets are larger than those of the mid-fifties but below
those of the mid-sixties, and the need has grown sharply. A substantial part
of this need is due to increases in the costs of basic research. It may still be
possible to advance basic science by flying a kite in a thunderstorm or by
peering through a homemade microscope, but it is much more likely that
advances will be made by use of powerful particle accelerators and electron
microscopes, operating with computer-controlled precision. Even at today's
prices, Franklin's kite, made of a "large thin silk handkerchief," would cost

only a few dollars, whereas construction of a powerful particle accelerator might cost several hundred million dollars. And yearly maintenance would require more millions. Franklin's letter to Peter Collinson describing his kite experiments took slightly over 400 words. Collection, storage, retrieval, and analysis of information collected in a single accelerator experiment requires extensive use of specialists and additional complex computer facilities. In brief, science now operates on an entirely different scale from that of earlier times. The growth of costs in basic research is not in itself an argument for whether or not more funds are needed, but it is important information that has to be considered.

The second point regarding basic research funds has to do with the adequacy of financing certain types of research. For example, in fiscal 1964 the social sciences received a total of about $33 million from the federal government. This was to be spread over the disciplines of cultural anthropology, economics, history, political science, sociology, and any other research undertaken primarily for the purpose of understanding group behavior. Even though most social-science projects are not as expensive as physical-science projects, the amount allotted is only a stale crust. The social scientists investigate problems that are crucial to individual, national, and international well-being. Future prospects for adequate funding of widespread social-science research are not very cheerful.

Probably the major reason why social sciences have been given so little attention is that social-science studies almost inevitably ruffle some Congressman's or special-interest group's sense of what is right and proper. Very few would understand or care about the implications if a new subatomic particle were found or the law of parity overthrown. An honest economist, sociologist, or political scientist can scarcely find a topic that does not outrage at least part of the populace. Those who are most likely to be outraged are defenders of the status quo, since serious research is unlikely to find that our social arrangements make this the best of all possible worlds.

On the other hand, it is sad to report that certain researchers in the social sciences have found ways to justify their role such that they do research which supports the "purse controllers" in Washington against their chief rivals for political leadership at local levels. This tends to assure these researchers of continued monetary support for their endeavors. How this works is spelled out in an interesting article by Alvin Gouldner (see Bibliography).

Even the basic research funds given to two specialized areas that are critical for the future is minimal; oceanography recently received about $28 million and agriculture about $23 million per year. (The present and growing pressures on our resources make both these allocations seem meager at best.) Our argument is not for restricting research in one area for the benefit of other areas. Few research areas are adequately financed. Only the degree of deficit differs.

The third point bearing on the adequacy of funds for basic research involves the support and training of future scientists. A substantial portion of the funds in any research program supports either graduate students as research assistants or junior scientists who work on particular projects. There is no substitute for apprenticeship training. The number of scientists trained for basic research depends on decisions made several years in advance, and the amount of funds allocated to basic research is one of the most important of such decisions.

Specific Distribution of Funds

Assuming a given amount of funds available for basic research, how are these awarded to specific investigators? We will consider only the distribution of federal funds, since the federal government is the largest single source. Practices in distribution of funds by institutional, private, and industrial sources vary too widely to be discussed meaningfully. However, it is interesting to note in passing that colleges and universities allocate about 84% of their own research and development funds to basic research, whereas, according to the 1966 survey, industry allocates only about 7% of its research and development funds to basic research. And this latter figure is probably overstated by at least 50% because of the glamour implicit in basic research.

The typical procedure in federal agencies is for the prospective investigator to submit a research proposal in some reasonably standardized form. A board of scientists evaluates the proposal and compares it with other proposals. The board of scientists is typically composed of men of some standing in their own discipline who serve only as government consultants. In general the system has worked well. However, one criticism is that these boards tend to be quite conservative. For example, funding of established scientists has been virtually assured while newly hooded Ph.D.s often have a rather difficult time unless they are personally known to some of the board members. The young Ph.D. may have a particularly difficult time if he proposes an innovation or departure from established thinking. Two possible solutions might be to earmark some money for young investigators or to give funds to various institutions for local distribution. There is a particular danger inherent in any system of research in which budgets are determined or funds allocated by administrators and/or committees of established scientists. This danger is that the direction of research will be where the money is. To some extent this does occur, and such a result is degrading to science and scientists. In addition, since administrators and established scientists tend to work out existing problems rather than create new problems or venture into new fields, this system is capable of producing sterility.

Inevitably there have been criticisms of federal research programs. Some of these criticisms are unquestionably justified. All programs need to be watched closely and criticized judiciously. Scientific programs are not exceptions. However, the critics ought to do their homework before launching their barbs. One of the best examples of tainted criticism was the lead editorial in a newspaper attacking a study of "mother-love" in monkeys as being an obvious waste of money. The wisdom of the editorial can be evaluated in the light of the fact that the scientist criticized, Dr. Harry Harlow, received a National Medal of Science award for this work—the federal government's highest award for distinguished achievement in science, mathematics, and engineering. Unfortunately, science has yet to discover the proper way to hybridize the "know-nothings" with the "say-nothings."

Two final points: (1) Three military-related groups—the Department of Defense, the National Aeronautical and Space Agency, and the Atomic Energy Commission—control approximately 88% of all federal research and development funds. No comment required. (2) Unfortunately, but contrary to popular belief, professors do not get rich on research grants. Generally, faculty members can draw only their regular salary or a portion of it through a grant. Research funds may relieve the faculty member of some other duty, may support him through the summer, or may lead to future advancement, but they don't directly increase his monthly paycheck.

Science and the Humanities

In recent years there has been considerable discussion of the relation between science and the humanities. C. P. Snow, the English scientist, novelist, and essayist, has presented the thesis that the intellectual community is divided into "two cultures," science and humanities. Like other sweeping generalizations, this view seems to be both true and untrue. It would be a trifle devious to argue that there are no differences in attitudes between science and the humanities at this point. But one would have to be a bit thick not to see that there are also a number of similarities. The idea of "two cultures" is an oversimplification, if for no other reason than that there are many cultures and each may be subdivided into a number of variations. It is worthwhile to ignore some of the finer points and consider the relation of science and the humanities.

The differences in attitudes are related in part to the different objectives of science and the humanities. In gross terms, one objective of science is to achieve precise and parsimonious statements about the structure and processes of the animate and inanimate world. Ideally, these statements allow us to describe, understand, and predict something about that world. As stated earlier, elegance or aesthetic appeal have their place in the world of the scientist, but these qualities can be expressed in terms of precision and parsimony. Einstein expressed the objective briefly: "Supreme purity, clarity, and certainty at the cost of completeness."

Some of the disciplines that fall into the general category of humanities are poetry, art, drama, music, literature, and speculative philosophy. Other disciplines, such as architecture (project housing excluded), history, linguistics, ethics, and cultural anthropology have a foot in this camp. In any event, a primary objective of the humanities is to enrich the life of the beholder by arousing some sensual experience, emotion, or feeling. Some of these feelings are quite complex and intricate, and need developing—an activity that requires a great deal of talent. Objectives such as precision and clarity are important in some cases; in others the effects may be achieved by complexity or a veiled hint. Granted that there are differences in objectives between science and the humanities, are there any necessary conflicts? The answer is an emphatic no. Ideally, the sciences and humanities should complement one another. Recall the plea in Chapter 5 for a broad education for scientists. This same plea can be extended to other disciplines as well. The devotee of the humanities who has no understanding of the general ideas of science is just as limited as the scientist for whom art forms arouse no emotion. The scientist doesn't have to be an expert (although it may help) in order to have his feelings aroused by Hamlet's soliloquy, the prison scene from Gounod's *Faust*, a line drawing by Picasso, or Richard Wright's *Black Boy*. By the same token, the general thesis of this book is that science is not a mystery that can be confided only to the elect.

Some of the differences between the humanities and sciences undoubtedly involve the personalities of individuals who are attracted to the different disciplines. There is some evidence that particular areas tend to attract and hold persons with certain personality characteristics. If this evidence is correct, differences among people also exist apart from particular disciplines—but these differences are only likely to be emphasized and enlarged by the discipline.

There are a number of ways in which the humanities and sciences are alike. One of the principal likenesses is the motivations of the professionals. There are probably two primary motivations for both groups. (1) Each participant has found his game, and (2) playing the game is important to him. In addition, both groups find in their games an escape from the crass, drab everyday world. It is interesting to note that their products make the world somewhat less crass and drab.

There is a very practical problem that has led to an emphasis of the division between humanities and sciences—the matter of financial support. Beginning with World War II there was substantial support for science, in great part due to the rather ignoble reason that science is necessary to war-making and international "one-upmanship." The humanities have not enjoyed this support. There are clear indications of envy and ill-feeling typical of the attitude of "have-nots" toward "haves." To be sure, the federal government has recently taken some timid steps in the right direction, but funding of the humanities is so minimal as to make the program only a sick joke. For example, the National Foundation on the Arts and Humanities spent only about $22.6 million for fiscal 1971—roughly the cost of one and one-half B52s. It is some sort of a commentary on our culture when we enthusiastically support the appropriations of billions for a war-making potential in order to preserve our "way of life," and we appropriate only a pittance to make that "way of life" richer, more meaningful, and in general more deserving of defense.

An argument is raised that, if artists, writers, and others in the humanities begin to receive government support they are likely to be controlled and directed. This threat just doesn't deserve to be taken too seriously. Science has remained relatively independent of governmental control even though primarily supported by governmental funds. Good scholars and artists tend to be very independent. The danger of government domination exists in all research and work, including the social sciences. But even under authoritarian governments such as Soviet Russia, Spain, and South Africa, intellectuals have proved very difficult to control.

Whatever the division or divisions between (and among) the sciences and humanities, there are two points worth considering. First, the amount of material in most disciplines is simply beyond the comprehension of any individual. In one particular field, approximately 24,000 books and articles were published in a year's time. Some other fields go far beyond this total. About 40,000 journals go to press regularly, and these journals contain over 24,000,000 pages annually. The point is clear: it is extremely difficult to be well-informed even in a single field. Universal scholarship like that possible during Aristotle's day does not occur. We cannot expect either the scientist or his counterpart in the humanities to be well-versed in many diverse fields, but the immensity and complexity of material is still no excuse for a lack of general knowledge. Second, there is great virtue in diverse approaches, even though diversity produces some confusion. The last great intellectual synthesis in Western Europe occurred in the Middle Ages and was enforced through the suppression of opposition. Intellectual stagnation was the price of uniformity. It seems obvious that pluralistic attitudes and objectives have the same virtues as a pluralistic society—variety, excitement, confusion, and the ability to change.

In spite of (or because of?) the differences in the funding of the humanities and sciences, the humanities do not seem likely to perish in the near future. Robert Nichols, program director of the National Merit Scholarship Corporation, surveyed the career decisions of National Merit semifinalists. The results are somewhat complex and show variations from specific field to field. Nichols summarizes the general situation thus: "The overall trend is that the interest of able students in physical sciences and engineering has been decreasing during the period covered by this study and that interest in the social sciences and humanities has been correspondingly increasing." Whatever the reasons behind this trend, let's hope that it signals an era of strength and innovation for both the humanities and the sciences.

Chapter 8
Science and Human Decisions

Science and Ethics

Our culture has a large number of values and ethical standards that are generally accepted. How do these relate to science and scientific attitudes? Some of the general principles in our society are the belief in free will and the belief that people are responsible for what they do. Are these beliefs scientifically meaningful? Do human behaviors follow laws in the same way that other physical phenomena do? Can they be understood through scientific methods? These issues are discussed in this chapter.

Terms such as "values" and "ethical standards" (or "ethics") are widely and loosely used. To avoid confusion, these terms need definition.

Not only do people behave, but they also feel and make categorical decisions about these feelings and behaviors. One kind of decision is whether a feeling is considered pleasant or unpleasant and whether it is approved of; such a decision, when made consciously, is a value judgment.

Since most feelings are associated with behavior, and since "value" also has the more general meaning of any score on a dimension, values tend to be confused with ethics, as we define the latter term. Another kind of decision is whether a behavior is considered good or bad; this kind of decision is considered an ethical judgment. Some behaviors, of course, are not given an ethical judgment. There are other ways to define values and ethics, but these will do for our purposes. If the ethical judgments made by members of a society about an individual's behavior are that it is good, he will usually be considered a moral or ethical person. The exception is that many behaviors are considered so bad that only one or a few instances of such a behavior will identify a person as unethical or immoral. High "goodness ratings" of behaviors are called "ethical standards."

Ethical standards are usually accompanied by some means of reward or punishment, which may range from social approval or disapproval to the use of physical force to enforce the standard. Since behaviors are accompanied by feelings, values always underlie ethical standards but values can exist without any ethical implications. For example, eating a very rare steak may be highly valued by one individual but rejected by another in favor of a very well-done steak. Acceptance or rejection would not ordinarily have any ethical implications. On the other hand, the knowing sale of tainted steaks would undoubtedly be considered unethical by both individuals. Still others might also condemn the eating of any beef on ethical grounds. For better or worse there are few, if any, human activities that would not be considered by some to have ethical implications. The individual determines whether a particular behavior or event is related to values and ethical standards.

Society tends to look at the relationship between science and human ethics in an inconsistent manner. On the one hand there is a great deal of fear that science falsely defines "good behavior"; on the other hand, some laymen expect science and the scientific methods to be the ultimate arbiter of ethics. Some people look to science for their ethical principles as others look to religion.

The relation between science and ethics is quite complex. Most of our attitudes, including those related to ethics and morality, have been affected by scientific thinking and discovery. Applied scientists have made many discoveries that are directly relevant to human behavior; but scientific concepts are not themselves behaviors and therefore are ethically neutral.

The ethical position of a scientist as a person is quite independent of his competence as a scientist. One can consider the behavior of a scientist according to two sets of criteria: first, his behaviors as a citizen, evaluated by his society; second, his behaviors as a scientist, evaluated by his scientific community. If he collects data in an unbiased manner, and if he can understand and interpret scientific achievements, he can be a good scientist no matter what his ethical position on other matters.

There are ethical considerations in a scientist's work, but ethical standards are not discovered by playing the game of science. The scientist may be able to say with scientific validity where a particular activity may lead, but he cannot use scientific principles to decide whether either the activity or its result is good. However, science can play a direct role in ethical determinations in at least two ways. First, given an ethical principle, scientists can determine some external conditions that are associated with behavior conforming to the principle, or conditions associated with behavior violating the principle. For example, let us assume that murder is bad. It is within the province of the social scientist to discover the conditions ordinarily associated with an increase in the number of murders. It has been shown that the uncontrolled availability of firearms is one such condition; therefore we can conclude that the uncontrolled availability of firearms is associated with consequences we have labeled "bad."

Second, scientists can establish what the consequences of certain behaviors actually are independent of any ethical considerations; on this basis, anyone can make his own ethical judgments. For example, children who are reared with limited stimulation tend to be permanently retarded in their physical, social, and intellectual development. By using all the information available, we may or may not decide that it is good to provide children with various levels of stimulation during their early years.

Some ethical judgments are made by specific individuals, some are made by particular groups, and some seem to be made by society at large. Many of these decisions are very difficult to make, and it takes a great deal of wisdom to make good decisions. The basis of this wisdom seems to be twofold: an appreciation of civilization and culture achieved through study of history, literature, art, and the other humanities on the one hand, and some scientific knowledge of the events associated with the problem on the other. This basis suggests that both the humanist and the scientist should have some knowledge of each other's fields and that the layman should be aware of both.

Although people use science to make ethical judgments, they do not make these judgments according to the principles of science. This statement needs some examples for clarification. There is scientific evidence that people who smoke cigarettes have a shorter life expectancy than those who do not. A smoker has a greater tendency to contract any number of diseases, including gastric ulcers and lung cancer. On the other hand, many people enjoy smoking. Also, there is some evidence that cigarette smoking relieves tension. And psychologists report that smoking is used by some people as a means for establishing social status and acceptance, and even as a symbol of virility or masculinity. Nowhere in this scientific description is there a statement about whether it is good or bad to smoke. There is simply a description of the actual consequences of smoking.

For years, scientists have been doing research in attempts to discover and understand the cause or causes of cancer. Many of these studies require procedures using live animals. These animals may be injected with cancer cells as part of the experiment and then are usually sacrificed in order that their tissues may be studied. A scientist can't use scientific methods to help him decide whether it is ethical to give these animals cancer. Rather, he uses ethical standards that he has attained elsewhere. He can use scientific goals as one of the bases for making his ethical judgments, but the decision about whether it is ethical to sacrifice animals is argued on other than scientific grounds. The argument favoring use of animals is sometimes defended on the grounds that new information or insights about the disease may be attained and that scientists will thus know more. This argument asserts that knowledge of disease is good, so it's ethical to acquire this knowledge, but scientists can't show that such knowledge is good. It is sometimes argued that this knowledge may lead to techniques to alleviate the disease in humans. This argument implies that disease is bad, but scientists can't demonstrate that disease is bad, either. The argument that sacrificing animals now to save the lives of humans later is often made. This argument suggests that human life is worth more than that of other animals, but again that assertion cannot be demonstrated. Sometimes the justification given for experimenting with animals is simply to gain more knowledge of the natural world. This justification and all the others stated here are generally accepted by our culture, but the real motive might simply be that some scientists enjoy watching animals suffer.

There is nothing in science stating that it is good to attempt to save human lives. Saving human lives seems to be a generally held value in most cultures of the world, but it is not in any sense scientifically derived. Nor is there any scientific derivation stating that it is not good to sacrifice animals. Some scientists feel no need to justify the sacrifice of animals under any conditions. Most hunters see no need to justify killing or crippling animals. Standards of scientists, like the standards of others, vary.

Consider another example. Scientists have discovered some of the causes of certain birth defects such as mongolism, and they can determine long before birth, with little or no risk to either fetus or mother, whether a fetus has such defects. But science does not say whether it is ethical to induce an abortion. The methods of science cannot specify any conditions under which abortion is ethical. Scientific methods can only lead us to understand the actual consequences of different actions. Look at the other side of this ethical dilemma. Is it ethical for a physician who knows that a fetus is deformed not to recommend abortion? The question is outside the realm of science, whichever way it is asked.

Should a scientist sacrifice a dog to find out what the effect of a drug is? Should a scientist build a hydrogen bomb? Should a scientist develop a chemical that in minute quantities can kill millions of people? Should a scientist develop a contraceptive pill? Science cannot answer these questions. The scientist can only say "If such and such is done, then this will follow." Ethical criteria from other sources are needed to decide what ought to be done.

In summary, if we ask the question "Do scientists playing the game of science have anything to say about ethics?" the answer is a resounding yes. Applied science, in particular, is important, because the applied scientist is the person who attempts to devise ways to achieve the ethical goals desired by individuals. Ethical goals have to be established in nonscientific ways, but once they are established, the scientist has the best opportunity to find out how to reach them. The proof that science is independent of any particular ethical system can be seen in the fact that a scientist can use science in an attempt to reach an amazing array of goals: to save a life or to take a life; to make someone fearful or to make him unafraid; to build a better bomb or a better food; a better mouse trap or a better mouse. Once the goal is established, scientific methods provide the best means for reaching them. But none of these goals can be defined as good or bad by scientific methods.

Values and Ethics within Science

The game of science normally has some generally agreed upon internal values. Some of these were previously discussed. For example, scientists tend to approve of knowledge and understanding. They think that it is good to understand an explanation or to see the relationship between two events, and they think that an individual should strive for understanding.

There is some disagreement among scientists about the breadth of the view that knowledge is good. The consensus among scientists is that knowledge and understanding are good in their own right and that individuals should strive for them. Some, however, feel that knowledge in and of itself is not important; it should lead somewhere. They say that only knowledge leading to practical consequences is good. All scientists, however, tend to put some positive value on attaining knowledge about certain aspects of the natural world.

Associated with the value of knowledge is a value placed on skepticism and tolerance. Since all scientific information and knowledge is tentative, scientists must claim to keep an open mind and be tolerant of those who disagree. This is a value that scientists preach quite regularly; but being human and involved in their beliefs, they don't always practice it as they should. Some scientists tend to evaluate opposing views with a less than open mind even while they are defending the concept of tolerance.

To say that a scientist believes scientific information to be tentative does not necessarily mean that he has no conviction in his beliefs. A scientist working within a paradigm is usually strongly committed to it. He often secretly believes his way of viewing the world is true, and his task is to prove it. But he has to realize that even though he is totally committed he may be mistaken.

Since knowledge of natural phenomena is their goal, scientists place a high value on accuracy in reporting data. The quickest way to be rejected by the scientific community is to be caught forging or deliberately misrepresenting data. All scientists make errors. Many factors are involved in the collection, analysis, and interpretation of data, any of which may go wrong, so errors are expected. But deliberate lying or forging of data can lead even a mild-mannered scientist to fury. Note that in the present case the underlying value is accuracy of reporting. The attitudes of the scientific community indicate that they have formed an ethical standard associated with accurate reporting. It is interesting to note that even though scientific enterprises would tend toward a high state of confusion if scientists did not report data accurately, that alone does not establish the ethical standard. The conclusion that science should not be confused is an ethical one and is not scientifically derived. There is no scientific basis for saying that confusion is good or bad. We might plead that a totally confused science cannot result in an increase in knowledge. The results of confusion can be based on evidence and logic and are therefore amenable to scientific investigation. Again, this points to the role of science in ethics. The consequences of a particular behavior can be investigated, but an absolute judgment that these consequences are good or bad cannot.

Even the few ethical standards we have attributed to scientists are not logically necessary. One need not hold them in order to be a scientist. It is possible for one to believe that the search for knowledge is bad and still search for knowledge. Someone may believe, for example, that the Bible should be taken literally and that it is wrong to challenge it; he may believe that the theory of evolution corrupts. Concurrent with these beliefs, he might carefully investigate the relations between animals, describe them quite accurately, and still call someone a sinner if he uses the data to support the idea of evolution. He may even consider himself a sinner whenever he considers the plausibility of the theory of evolution. It is not the ethical standards he holds that make him a scientist; it is the procedures he follows (although inevitably these procedures are conditioned by his attitudes). People who engage in certain activities tend to justify these activities to themselves; thus, their ethical standards tend to coincide with their activities. It is because of these relations between behaviors and ethics that the ethics of scientists can be consistently identified.

Free Will and Determinism

There is another way that behaviors and ethics are related—one that affects some of the bases of our society. The relationship between behaviors and ethics includes such diverse topics as morality, sin, reward and punishment, compulsion, natural law, causality, responsibility, knowledge, fatalism, predestination, and especially the central issues of free will and determinism. Specifically, what a person thinks is true about the determinants of his behavior almost necessarily affects his evaluation of different behaviors.

Do people have free will? Do we have real options? Are we responsible for our actions, or are these actions merely a necessary outcome of natural law? If our behaviors are predetermined, could there be any reasons to try to change them? What about our penal system? If behaviors are determined, why do we punish those who violate our laws? If they could not help it, are they responsible for their actions? The same question applies to positive accomplishment. Why reward Willie Mays for hitting home runs in baseball if his skill was predetermined and he couldn't help it? If we should punish and reward, and if we are responsible for our own actions, aren't behavioral and social sciences a misnomer? Are they really sciences?

No one can answer all of these questions in a totally convincing way. Philosophers have been writing about such problems for centuries and are very good at disagreeing with one another. Probably no scientist lets these questions interfere with his research. Scientists must take sides on some of these issues—at least they must act as though they do. They have to assume that human behavior is determined. In other words, the scientist has to believe that he can find out things about people and situations that limit the options of an individual in a situation. He may further believe that he can gain information about the individual that will allow him to predict with great accuracy what the individual will do. And many scientists can demonstrate that modifying a situation in a particular way will result in the individual's behaving in a certain way.

Statements such as those in the previous paragraph put the social scientist into conflict with certain accepted values and beliefs. For example, most people believe that they have free will. Much of our legal code and much religious dogma is based on the assumption of free will. Do social scientists believe that people are responsible for their actions? Should good actions be rewarded and bad ones punished? In order to attempt to answer these and other questions mentioned in this section, we must first analyze what the terms "free will" and "determinism" mean.

The most obvious meaning of free will is that you can do whatever you want to do. If you want to sleep, you go to sleep. If you don't want to go to work, you stay home. If you want to go to a ball game, you can go.

It is immediately clear that there are severe limits to one's choices. The ordinary individual can't walk through a wall, run 100 yards in 9 seconds, fly through the air, end a war, win the National Open Golf Tournament, or score 190 on an I.Q. test. There are obvious physical and mental limits beyond which most of us can't go.

Now we can talk about *determinism*—the idea that all events are functionally related, at least on a probabilistic basis, to antecedent events. If certain physical and psychological laws determine the limits beyond which free will cannot operate, it may be said that free will and determinism can exist together and are not necessarily opposed to each other. But since it is unlikely that most people would seriously and rationally will to do all the things mentioned above, there must be other criteria to identify what free will means. Free will is usually claimed to exist within the limits of physical possibility. It is said to exist among reasonable and plausible choices. Thus, a slightly more sophisticated definition of free will is that it is the ability to decide which of several possible things to do, together with the ability to do the thing chosen.

According to this definition, free will is the relationship between making a decision and following through on it, within the limits of physical possibility. This definition is not so grandiose as the first, but it has interesting related considerations. The most striking consideration is that a limited definition of free will such as this is acceptable to a great many deterministically oriented behavioral and social scientists. These social scientists will hasten to add that believing a person can make a decision and follow through on it in no way contradicts a belief in determinism. The belief that free will and determinism are in conflict shows a lack of understanding of determinism.

The concept of determinism is essential to an understanding of the social scientist's position, yet it is usually misunderstood. Many people confuse determinism with fatalism or compulsion. All that a scientist who is a determinist wishes to do is explain and predict behavior. He thinks that the more he knows about a person and a situation the better he will be able to predict what the individual will do. A scientist may even set up situations under which certain behaviors will occur so that he may improve his ability to describe and understand behavior regardless of his attitude or interpretation of free will.

A person's ability to act on his decisions does not rule out determinism. A determinist may simply try to understand what the person will want to do. Any person who has free will demonstrates at least some limited amount of determinism: he has to be able to predict (determine) what he will do after he decides to do it. If there were no determinism whatever, all behavior would be random.

Fatalism and Scientific Determinism

What about fatalism? How does it differ from determinism? A fatalist is necessarily a determinist, but a determinist is not necessarily a fatalist. A fatalist believes that certain results are destined to happen no matter what a person does. Such sayings as "A person will die when he is fated to die" and "If a bullet has your name on it, it will hit you" are examples of fatalistic beliefs. People who predict dire (or pleasant) consequences on some future date also tend to use fatalistic ideas. These ideas are contrary to scientific determinism. A scientific determinist believes that there are functional relations between variables, and he uses his knowledge of these to predict the future. His predictions are much more mundane. He recognizes that there are unknown aspects of the environment that will affect the outcome. A scientific determinist will say such things as "If you are in a battle and stick your head out of a foxhole, you are more likely to die than if you do not stick your head out of the foxhole." Or "If you have a habit of driving at 80 miles an hour in the city while intoxicated, you are likely to have an accident." Or "Bob Jones, who scored 110 on an I.Q. test and wants to be a theoretical physicist, is in for a rude awakening." The point is that the determinist makes small predictions based on some specific relations among those things that he knows about, whereas the fatalist makes major predictions based on some mysterious power of prognosis.

Let's look at a short hypothetical example. Charlie Green is a very rigid individual. He eats lunch at the same cafe at the same time every day. At precisely 12:03 each day he turns the corner at Second and Main Streets on his way to the cafe. On February 2 a new company is moving into an office on the third floor of the building at Second and Main. They start raising a 300 pound safe, using a block and tackle, at precisely 12:00. It just so happens that the rope is fraying and will be able to hold a 300 pound weight for only three minutes. Now a fatalist watching this scene might say that poor Mr. Green is doomed to be smashed. A determinist, on the other hand, might say that if everything remains the same with relation to Mr. Green, he is likely to round the corner and be killed. The determinist also thinks he can be of help, so he tells Green that a safe is being lifted and that it is dangerous to walk under it. Mr. Green says "Thank you," looks up (for the first time in five years), sees the safe, and detours into the street as the safe falls harmlessly. One of the determinants of Green's behavior is the verbal information he receives. A deterministic social scientist would be able to predict (determine) Green's safety, based on the information Green received. The safe was not fated to land on Green's head. Only if no one informed anyone else of the situation would it happen. The determinist allows for adjustments in a changeable situation. The fatalist simply says afterwards, "Well, the safe didn't have Green's name on it."

What does all this have to do with free will? It seems to suggest that at this level of analysis free will and determinism are not incompatible. Mr. Green had the free will either to walk under the safe or to detour into the street. His decision rested partly on his knowledge of the situation. By studying Green beforehand, the social scientist can determine Green's choice of action. The study might reveal that Green is extremely suicidal and wishes his death to appear accidental, in which case the scientist may predict that Green would ignore the advice.

An individual with free will has the ability to consider the information available, use it to make a decision, and act accordingly. A determinist who understands the laws of behavior can predict with considerable accuracy how the individual will interpret information, what decision he will make, and how he will act. A scientific determinist cannot predict with complete accuracy, however, because he does not know all the particular events an individual will encounter. If he did know certain events an individual would encounter, by informing the individual the scientist could affect a change in his behavior. In other words, behaviors are not fated to occur. The individual's behavior is partially based on the information he has.

There are times, according to this deterministic position, when an individual does not have free will. There are some occasions when, for one reason or another, an individual cannot do what he consciously wishes to do. An example of the lack of free will is compulsive behavior. Behavior is considered to be compulsive if the individual feels compelled to perform certain behaviors when he doesn't wish to. There are many examples of compulsive behavior, both normal and abnormal. For example, there is what is known as the peanut phenomenon; if a person decides to eat a single peanut he may find that he has a compulsion to eat another. Scratching a mosquito bite is another compulsion. Some people have more serious compulsions that may keep them from doing what they wish to do or lead them to do things they don't wish to do.

The peanut phenomenon and scratching are examples of *internal* compulsions. There are also *external* compulsions, which occur when an individual wishes to do something and under normal circumstances could accomplish his aims but is constrained externally. An example would be someone who wants to go on a picnic but is locked in a jail cell. Closely related to this kind of constraint are situations in which an individual is forced to do something by threat, such as a bank teller being forced under gunpoint to give money to a robber. Thus, an individual is thought *not* to have free will when either by internal or external pressure he is compelled to do something which he does not desire to do.

Considering the use of the terms "determinism," "compulsion," and "free will," as we have defined them, we can conclude that the real opposite of free will is compulsion, not determinism. There must be some determinism for an individual to have free will, since he has to be able to choose from among predictable behaviors. A knowledgeable scientist should be able to predict the behavior of people whether they are exhibiting free will or not. As the sciences advance, predictions of behavior will become more accurate and broader in scope because the known functional relations between variables and the theories of behavior will be more closely matched to the events.

Free Will and Scientific Law

One of the reasons people tend to think that the game of science cannot be applied to human behavior is because they misunderstand scientific law. They think that scientific laws are like political laws: an individual is forced to follow them (or perhaps will be punished). But scientific laws are not prescriptive—that is, they don't say how people or things ought to act. Scientific laws are descriptive. They describe how people and things do act. If people don't obey a law proposed by a social scientist (that is, act as the law says they act) the people are not punished; rather, the law is seen to be invalid. A scientific law does not compel behavior, it simply describes it.

A scientist in any field of endeavor attempts to use scientific methods to find out what actually happens under different conditions. By investigating the conditions that lead to different events, scientists identify variables and try to invent theories from which valid laws of behaviors can be derived. It is possible that, as the behavioral and social sciences develop, scientists will be able to describe conditions that tend to ensure that an individual has free will. In other words, they will describe scientific laws that determine free will, and they will describe ways to manipulate variables so that compulsions disappear. Scientists may also describe the variables and scientific laws that determine the choice the individual will make.

We probably will never know all the scientific laws that determine human behavior (or the behavior of atoms), but since scientific laws are basically descriptions of the conditions under which behavior takes place, it makes no sense to say that behavior does not follow the laws of science.

To summarize, a social scientist may be able to predict behavior and thus demonstrate determinism. Much of the time an individual exhibits free will. People do not always have free will because there are occasions in which they are compelled to act regardless of their will. Such compulsions may be due to either internal or external forces. The scientist who knows (1) the laws of behavior in general, (2) the individual in particular, and (3) the current circumstances can predict behavior whether the individual has free will or not. But scientists will never be able to predict all behavior exactly, because they can never know every possible event that may occur.

Reward, Punishment, and Responsibility

If all behavior follows scientific laws, what is the function of reward and punishment? Are individuals responsible for their actions?

The first question has a scientific answer. Reward and punishment have an effect on behavior, so the laws of their relation to behavior can be scientifically investigated. Many psychologists think that reward and punishment are two of the major determinants of behavior, although their effect is a complex one. There is a tendency for individuals who are rewarded for a certain behavior to repeat the behavior or at least the same kind of behavior. The effect of punishment is not so clear; punishment does seem to have at least a short-term inhibiting effect on the behavior punished, and under some conditions these inhibiting effects are very long lasting. In any event, scientific evidence makes it quite clear that reward and punishment and their interpretation have a considerable effect on other behavior. Social scientists spend a great deal of time investigating these very important variables.

Ethical judgments are at least partially affected by the functional relationships between reward, punishment, and behavior. First of all, decisions must be made concerning which behaviors are to be rewarded and which punished. Such decisions influence immediate behavior; but more importantly, they have permanent effects on the individual's behavior.

The function of attribution of responsibility is not so clear. Social scientists have found that people do attribute responsibility both to themselves and to others (and sometimes even to inanimate objects), and they have also found that behavior is affected by the belief that people are responsible for their actions. There are times when people do not hold others responsible for actions; for example, small children and those who are legally incompetent are usually not so held. These beliefs, the conditions leading to them, and the consequences of these beliefs are beginning to be scientifically investigated, but not much is currently known. Whether people *really* are ultimately responsible for their actions is a question that cannot be answered scientifically. There are always antecedent events that determine behavior, but the feeling of responsibility (which has antecedent conditions determining the feeling) in turn is one of the determinants of behavior. Whether a person is considered responsible depends on who is doing the considering and whom he considers.

The fact that behavior is determined is no reason to shirk responsibility or to give up work with the assumption that since behavior is determined there is nothing you can do. Two major determinants of what you can do are your feelings and your beliefs of what you can do. And in spite of the fact that behavior is determined, no one knows for certain everything that will happen. If one could predict every event, one could effectively control all events.

Chapter 9
The Scope of
Science

The object of this chapter is to organize some of the matters we discussed earlier into a context that will clarify two important aspects of science—its breadth and its limits.

The Breadth of Science

Although we have concluded that scientific methods are broadly applicable, we discussed the current breadth of science only briefly. The number of different paradigms and different kinds of events recently investigated by scientists is very large, and the number of different bodies of knowledge that have been built up and the amount of information that has been gathered is truly remarkable. Until funding was recently cut and applied work emphasized, research was proceeding at an ever-increasing pace. At present, basic research appears to have leveled off or possibly declined. Even so, the number of scientists continues to increase.

Technology has developed to such an extent that, with the assistance of computers, scientists can collect and analyze much more data than was possible two short decades ago. It has been estimated that the total amount of man's knowledge has been doubling approximately every seven years. Because of the vast expansion of knowledge, many scientists feel that they have to specialize narrowly to keep up in a given field. Specialization has two unfortunate limitations: individuals interested in more than one field are rarely able to explore these different interests; and scientists are often unaware of work from other fields that is relevant to their own work. In spite of these drawbacks to overspecialization, most scientists must do research on relatively narrow topics in order to make a meaningful contribution. As these topics accumulate their own bodies of data, theories, and methods, the scientist has to specialize even further.

New sciences often develop when a research area is seen to lie partly in each of two sciences. One current problem of many universities is that formal departmental structure may prevent students interested in one of these new sciences from getting the best education available. At present, such students must elect associated courses in all the relevant departments.

In the last few years the social sciences have come of age. Social scientists no longer have to consider themselves second-class citizens among other scientists. Because of the growing bodies of knowledge and new techniques for gathering information, sociologists, economists, political scientists, psychologists in some areas, and historians can take their place along with the natural scientists. The social scientists have difficulty doing certain experiments—many of their problems cannot be isolated in the laboratory, and ethical considerations also prevent some research—so they have developed some of their own techniques. In certain instances, applied and basic science merge into what has been called "action research." The value system of the scientist combines with the search for understanding; the research is done in a natural setting with the theoretical expectation that the results of the research will be goals that the scientist wishes to achieve in society.

Considering the kinds of problems social scientists have, a compromise between theoretical and ethical considerations seems a valid resolution of the conflicts. However, isolation of variables in the laboratory, which allows unconfounded observation, is more likely to lead to greater understanding of the processes involved and to better description of relevant functional relations between the variables. Therefore, experimental investigations should be used whenever feasible.

Although limited in the experimental procedures available, certain social scientists in the last few years have been developing some formal theoretical approaches, such as game theory and computer simulation, which may lead to major breakthroughs if they have not already done so.

One social science that has shown remarkable productivity and popularity in the last few years is actually one of the oldest sciences of man: linguistics. Even ancient civilizations studied their languages and tried to formalize some of the facts about them. But with recent interest in generating sentences, with the suggested possibility of computer translation, and with new techniques of analyzing speech, new problems with new attempts at solution have emerged. Many scientific fields are appearing with linguistics at one focus; biological linguistics, psycholinguistics, sociolinguistics, and mathematical or computational linguistics are four of them.

Just about everyone is aware of the recent major accomplishments of the sciences associated with medicine. One of these, genetics, which includes an analysis of the details of genetic inheritance and corollary physiological functions, portends many ethical problems when we learn to predict and control human heredity, but it is still a fascinating area of scientific study. Other areas are equally exciting. For example, the theoretical and practical problems associated with organ transplantation are of great importance in understanding human physiology.

Among the many different and interesting problems in the many sciences, each of which often seems more important than the others, is one science that is extremely new and relevant to just about every other modern discipline: computer science. Associated with computer science are some general problems relating to symbol manipulation, problem solving, artificial intelligence, the monitoring and modifying of systems (any systems), the rapid analysis of data, and the study of automata (machines). The computer itself is a scientific and technological discovery of tremendous importance. It has made many previously impossible activities routine. Whereas a man may be able to make a few thousand operations per day, a computer may make one hundred million operations per second and make decisions at the same time. It receives, processes, and acts on information in billionths of a second, whereas man's fastest simple actions take about one tenth of a second. To understand and be able to use this tool wisely requires the special attention of some scientists. Computer science, however, is not simply the study of a machine. The development of the computer has pointed to natural problems of considerable interest, such as the differences among and between natural and artificial languages, the theory of problem solving, the theory of automata, memory systems, information processing, and perceptual systems.

Scientists in all fields as well as a variety of individuals such as bankers and production line workers will find the computer an important adjunct. Scientists and nonscientists alike should have at least a nodding acquaintance with it.

Science, as it develops and advances, is being carried into a myriad of different areas, and the number of areas and paradigms is multiplying rapidly. Science not only has extended into many new realms of the natural world but is extending into many aspects of the man-made world, such as computers, business organizations, social institutions, and plastics. All conceivable topics are not discussed in science, but very few topics can automatically be excluded.

The Limits of Science

As we have seen throughout this book, many important aspects of the way we live can be and are affected by science. The environment in which we live has been radically changed by science and technology. Our attitudes, beliefs, and even ethical principles are influenced by scientific knowledge. However, even though science has these widespread influences and affects all aspects of our life, it has important limitations. We have looked at a few problems for which science has no answers. Now we should discuss the limits of science as a topic in itself.

The most important limitations of science are an integral part of its methods. There are three ways these limitations can be clearly seen: first, both data and statements about that data are only probabilistic; second, there are both technical and theoretical limits to the accuracy of measurement; third, whatever the accuracy of our measures, we cannot measure all aspects of a given event. As stated in an earlier chapter, information about the real world ultimately has to be known through induction, and nothing that is known by induction is known for certain. There is no way one can gain certain knowledge about the real world.

In the end, scientific statements stand or fall by induction, and it is inductive methods that have led to, and provided tests for, atomic fission and fusion theory, theories of natural selection, and computers. Scientists may or may not believe that their statements will be shown to be ultimately true, but if one accepts the concept of induction, scientific statements can be demonstrated to be reasonable.

Another limit of science concerns the accuracy of measurement. This accuracy is limited in a number of ways. To begin with, there are limits on measurement due to technology. Intelligence and socioeconomic status are important variables, but we can't measure them accurately; we can't measure the weight of a gas to a great degree of accuracy; we can measure the speed of light only within limits; and so on. We can't measure anything completely accurately, and we never will be able to, if our current conceptions of physics are accurate. There is a principle in physics, known as the Heisenberg uncertainty principle, which leads to the conclusion that there are limits to the accuracy of our measures. When subatomic particles are involved, there is no way to measure simultaneously both their velocity and their location. Thus, one can't accurately predict a particle's future location and velocity. Furthermore, the apparatus used to measure subatomic particles affects their location and velocity, so the measuring instrument affects the object measured. In certain psychological and sociological measurements, there are different but parallel effects. Sometimes the test a person takes or an interview he experiences affects his behavior. One may not then be able to identify what his performance would have been had he not taken the test. One other limit on the accuracy of measurement should be mentioned. Certain measurements take time to collect, particularly in the social sciences—for example, gross national product, or data on absenteeism. If the data are collected in a dynamic system, then the data do not represent the current situation by the time they are collected.

We can never know everything in science. Any specific event in the real world can be analyzed or categorized in an unlimited number of ways. Each of these has a very large number of variables affecting it. Since there are essentially an unlimited number of events, and only a small number of scientists with a limited amount of time and facilities, so there will always be a great deal left to be investigated. The known world is incomplete and will remain so. We will never understand more than a small part of it. And since the ability of any scientist to conceptualize is limited, and the number of aspects of events unlimited, conceptualizations have to be incomplete.

We are limited in what we know of some events because, although we are attempting direct measurement, there is a limit to the accuracy of our measurement. Our knowledge of other events is further limited because there is no way to make direct measurements. In these cases, we have to infer the state of these events from indirect measurements. Consider human values; they are what make life worthwhile (the fact that we think life is worthwhile is a value judgment). On what basis do we decide that something is good? Beautiful? Fun? Interesting? A psychologist may be able to answer each of these questions. He may be able to find out what we *think* is good, or beautiful, or fun, or interesting, and he may eventually be able to tell us why we think so. He will come to these conclusions by indirect measurement. But in *no* instance will he be able to tell us whether something is *really* good, beautiful, fun, or interesting. You may object, saying that something may *be* really good or really beautiful, that something is *automatically* fun or interesting if we enjoy it. You may argue that fun, by its very nature, cannot be analyzed any further. Your argument may or may not be valid. It is irrelevant. The point is that a scientist can do the same kind of investigation with any of these values. Going beyond these limits to say whether something is *really* good or *really* beautiful is not within the realm of science.

These topics undoubtedly need to be analyzed, and philosophers do, and should, investigate the question of absolute good, just as psychologists and other social scientists investigate what people think and believe about goodness. But there are limits to any of these analyses. A philosophical analysis may show what logically follows from certain assumptions about values, and these results may then be evaluated, but nowhere will the analysis state "something is good" with certainty. A theological, religious, or mystical source may indeed state that "something is good" and one may feel certain of it, but there is no scientific evidence to justify the feeling of certainty. All that anyone can do is to decide on some intuitive basis that something is good and (perhaps) act accordingly. To help his intuitions and to help him make these decisions about goodness, a scientist may use the knowledge of the consequences of certain activities. That is, he may evaluate the consequences as well as the activities.

Exactly the same arguments hold for other values; scientists can only find out what people believe and the consequences of such beliefs; they cannot demonstrate the truth of such beliefs.

The last "limit of science" we will discuss falls under the general heading of "metaphysics." These questions are about reality directly; not: is it really good, or really beautiful? but: what *is* it really? What are electrons? What is a table really made of? Do minds really exist? Do people have immaterial souls? Does magnetism exist? These questions can be most exasperating. And they cannot be answered. A scientist starts with a set of undefined terms. He may think that they are real, and he may give them some properties, but he can't tell you what they really are. He may not even understand the question. A scientist attempts to confirm a theory in terms of observations. He can specify the conditions for observation and he can describe the properties of the events observed. He can relate the events to other events in a meaningful way. Those are the limits of his ability. Due at least in part to the limitations of science, some people turn to other areas in an attempt to get true and certain information. They may turn to authority, intuition, superstition, psychedelic drugs, or religion. These sources may give an individual the feeling of certainty, but there is no logical reason why a feeling of certainty is any more likely to be true than a belief without the accompanying feeling. Some data even suggest that feelings of certainty are often accompanied by more errors than feelings of uncertainty. Any system stating that something is true by fiat, authority, or vision cannot demonstrate its truth. A person has no reason to accept it as true whether he is certain that it is or is not.

In conclusion, it can be seen that there is no part of the observable world that is automatically immune to scientific investigation. Even those personal aspects of human existence that "feel" immune to science can be studied scientifically. There are no natural events that negate such study. Limitations, where they exist, refer either to measurement limitations, to ultimate solutions, or to values. We all hope that the ethical principles followed by scientists are those leading to "life, liberty, and the pursuit of happiness," and we also all hope that scientists consider both their actions and the consequences of their actions in terms of man's welfare. But since ethical judgments ultimately rest on an ultimate value, they are basically immune to scientific investigation. Only the events as such can be understood scientifically.

A Few Concluding Remarks

Four principal themes have been followed in this book. Each one appears in several sections and perhaps deserves re-emphasis.

Science has been viewed as a game not because it is trivial or frivolous, but because it has many of the attractive elements associated with games.

Monkeys will work long periods of time solving puzzles, and most scientists are fascinated by intellectual puzzles. Whether monkeys and scientists should be grouped in this way is a matter for speculation, but the evidence is clear that working on puzzles is intrinsically rewarding to both groups.

Beginning in Chapter 2, we emphasized the importance of attitudes. There are many possible ways of analyzing or examining any object or event. The attitudes and values of scientists set the boundaries of what is an acceptable statement or item of evidence within science. The emphasis is on objective and verifiable evidence and statements that are clear, logical, and at some point related to or confirmable by production of evidence.

Discussions of science have sometimes revolved around the scientific method. We take the view that, while there are methods within sciences, there is no single scientific method. This is not to deny the importance of methods and procedures, but rather to stress that methods and procedures are related to criteria for acceptable hypotheses and evidence. For example, social behavior of nonhuman primates is a topic of current interest for ethologists and psychologists. Their research strategies and methods differ widely. Ethologists typically favor field studies in which animals are studied in their natural habitat. Psychologists are more likely to study such behaviors in a laboratory setting. Each method has advantages and disadvantages. Whatever the method used, the common objective is to secure reliable information. Methods that do not lead to reliable information would be rejected by both ethologists and psychologists.

In several sections and particularly in the last two chapters, we have attempted to outline both the scope and limitations of science. Although there are limitations to science, we must not overlook the fact that the scope of science is extremely broad and far-reaching. Complex but understandable. Fallible, but capable of self-correction. Restricted, but with fantastic capability for growth. Imperfect, but struggling. And, of course, it's fun.

Annotated Bibliography

The books cited below were the most direct references used in writing this book. We did not usually reference them directly in the text, since we did not usually present specific information having only a single source, and we hoped to have the book flow more by omitting footnotes. We annotated the bibliography to help you find books and articles about science that may interest you.

Baker, R. A. *A Stress Analysis of a Strapless Evening Gown.* Englewood Cliffs, New Jersey: Prentice-Hall, Inc., 1963. A collection of humorous and satirical views of science, mostly by practicing scientists.

Barnett, L. *The Universe and Dr. Einstein.* New York: New American Library, Inc., 1957. A popular presentation of the view of the world according to modern physics; well-written and clear.

Blackett, P. M. S. "The Ever Widening Gap," *Science,* 155 (1967), 959–964.

Boring, E. G. *A History of Experimental Psychology* (2nd edition). New York: Appleton-Century-Crofts, Inc., 1950. A fairly difficult and detailed but well-written history. A good discussion of the concepts and people who led to modern psychology.

Carnap, R. *Philosophical Foundations of Physics.* New York: Basic Books, Inc., 1966. A cogent presentation of philosophy of science by an eminent philosopher. The book was edited by the well-known writer Martin Gardner.

Cartter, A. M. Scientific Manpower for 1970–1985. *Science,* 172 (1971), 132–140. A very straightforward, if somewhat alarming, analysis of the long-range future of scientists. Incidentally, Cartter predicted the current oversupply of Ph.D.s when others were crying shortage.

Coler, M. A. (ed.), *Essays on Creativity in the Sciences.* New York: New York University Press, 1963. Considers creativity in a wide variety of views and contexts. Quality of the essays varies considerably.

Conant, J. B. (ed.), *Harvard Case Histories in Experimental Science* (2 volumes). Cambridge, Mass.: Harvard University Press, 1957. Details of different specific episodes in the development of the natural sciences.

Conant, J. B. *Science and Common Sense.* New Haven: Yale University Press, 1951. A very readable and clear attempt to give the non-scientist an understanding of science.

Condon, E. U., and Odishaw, H. *Handbook of Physics* (2nd edition). New York: McGraw-Hill Book Company, Inc., 1967. A difficult general text of many aspects of current physics.

Crompton, J. *The Life of the Spider.* New York: The New American Library, Inc., 1954. A light and fascinating informal view of spiders and people who study them.

Darwin, C. *Origin of Species.* New York: Modern Library, Inc. What can we say? The "Historical Sketch" alone is worth the price of the book.

Einstein, A. *Essays in Science.* New York: Philosophical Library, Inc., 1934. Translation of a number of short essays. Most of these are readable with little or no scientific background.

Einstein, A. *Ideas and Opinions.* New York: Crown, 1954. More short and mostly readable comments and essays.

Feigl, H., and Brodbeck, M. (eds.), *Readings in the Philosophy of Science.* New York: Appleton-Century-Crofts, Inc., 1953. A large book of readings covering many aspects of philosophy of science ranging from relatively easy to quite technical and difficult.

Gardner, M. *Fads and Fallacies in the Name of Science* (2nd edition). New York: Dover Publications, Inc., 1957. A brief look at pseudo-science.

Gouldner, A. W. The Sociologist as Partisan: Sociology and the Welfare State. *American Sociologist*, May 1968, 103–115. A readable discussion of the interaction of values held by scientists and their science.

Hagstrom, W. O. *The Scientific Community.* New York: Basic Books, Inc., 1965. A survey and analysis of social influences, particularly those from within the scientific community, on the conduct of research.

Hansen, N. R. *Patterns of Discovery.* New York: Cambridge University Press, 1958. An interesting and readable presentation of some aspects of the paradigmatic approach to science from a slightly different orientation. It is our major source on Kepler's research.

Hildebrand, J. H. *Science in the Making.* New York: Columbia University Press, 1957. A delightful little book by a practicing scientist. To be commended for insight (since we seem to be in agreement on most matters).

Hodgeman, C. D. *Handbook of Chemistry and Physics*. Cleveland: Chemical Rubber Publishing Co. The standard handbook of many different kinds of mathematical, chemical, and physical tables.

Huff, D. *How to Lie with Statistics*. New York: W. W. Norton & Company, Inc., 1954. A humorous little book that presents ways people can misrepresent data to give impressions that are not valid. It tells what to look for in evaluating reports of data.

Hyman, R. *The Nature of Psychological Inquiry*. Englewood Cliffs, New Jersey: Prentice-Hall, Inc., 1964. This is a book on scientific methodology, written from a scientific paradigm point of view. All of the examples and details are taken from psychology.

Kohler, W. *Gestalt Psychology*. New York: Liveright Publishing Corporation, 1947. Max Wertheimer, the founder of Gestalt psychology, never wrote a survey of it. This lucid book was written by his colleague and co-founder.

Kovacs, E. *Biochemistry of Poliomyelitis Viruses*. New York: Macmillan, 1964. Used as the primary basis of our short history of virology. Very interesting, but many parts of the book require special knowledge.

Kuhn, T. *The Structure of Scientific Revolutions*. Chicago: University of Chicago Press, 1962. The book that got us started. An essay in the history of science, it contains a clear presentation of the paradigmatic character of science.

Ladd, E. C., and Lipset, S. M. Politics of Academic Natural Scientists and Engineers. *Science*, 176 (1972), 1097–1100. A survey of the general political orientation of a wide variety of scientists. The authors find clear differences in various fields, also differences related to the individual's standing within his field. Extremely interesting, with some surprises.

Langmuir, I. "Pathological Science." Colloquium at the Knolls Research Laboratory, December 18, 1953. This amusing and tragic transcript was made by R. N. Hall from a disc recording furnished by the Manuscript Division of the Library of Congress. Langmuir reports a series of findings, primarily by physicists, in which they found things that were never there. Interestingly, some of these findings generated hundreds of supporting studies.

Lecky, W. E. H. *The Rise and Influence of Rationalism in Europe*. New York: George Braziller, Inc., 1955. Originally published in 1865, this is a fascinating, if somewhat repetitious, account of the development of modern attitudes. The opening discussion in Chapter 2 was inspired by Lecky.

Margenau, H., and Bergamini, D. *The Scientist*. New York: Time Inc., 1964. Includes a wide variety of topics and illustrations presented at a popular level.

Medvedev, Z. A. *The Rise and Fall of T. D. Lysenko*. I. Michael Lerner, trans. New York: Columbia University Press, 1969. The sad and inspiring story of Russian scientists' struggle against brute force and dictated science. At his trial, N. I. Vanilov declared, "We shall go to the stake, we shall burn, but we shall not renounce our convictions."

Mencken, H. L. *Prejudices: A Selection*. New York: Vintage, 1958. Witty, barbed essays on a number of topics. We share a number of Mencken's prejudices.

National Science Foundation. *Reviews of Data on Science Resources* and *Surveys of Science Resources*. These and other publications of the National Science Foundation contain a wealth of material on manpower, financing, distribution, employment, etc. within the scientific community.

Obler, P. G., and Estria, H. A. (eds.). *The New Scientist*. Garden City, New York: Doubleday & Company, Inc., 1962. A collection of essays about scientists, their personalities, values and relations to the world.

Ravetz, J. *Scientific Knowledge and Its Social Problems*. Oxford, England: Oxford University Press, 1971. Contains a section on critical science reprinted in *New Scientist* and *Intellectual Digest*. Proposes that substantial scientific effort be directed toward examining the effects of scientific and technological developments.

Rhine, J. B. *New World of the Mind*. New York: Apollo Editions, 1953.

Rhine, L. E. *ESP in Life and Lab: Tracing Hidden Channels*. New York: The Macmillan Company, 1967. Whether you happen to be interested in this sort of thing or are just idly curious, these books contain interesting discussions of ESP from the viewpoint of ESP researchers.

Roe, A. *The Making of a Scientist*. New York: Dodd, Mead & Company, Inc., 1953. This author has done more than any other single individual to study the scientist as a person. Other monographs and articles by this author are recommended.

Russell, B. *History of Western Philosophy*. New York: Simon and

Schuster, Inc., 1945. Undoubtedly the best written history of philosophy around, this is an articulate and sometimes humorous book by a great man. Since it attempts to do so much in a limited space it is superficial in spots, but it is still an excellent general introduction to philosophy.

Sachs, M. A. Resolution of the Clock Paradox. *Physics Today*, September 1971, 23–29. See also letters in response to this article in *Physics Today*, January 1972, 11–17. An interesting discussion of a current controversy in physics.

Schwartz, G., and Bishop, P. W. *Moments of Discovery* (2 volumes). New York: Basic Books, Inc., 1958. Short sketches of the life and activities of a wide variety of scientists beginning with Hippocrates. Short selections of their translated works are of particular interest.

Science. Published by the American Association for the Advancement of Science. Presents articles from many fields and on a wide range of levels. Strongly recommended.

Scientific American. Published monthly by Scientific American, Inc., New York. Each issue contains a variety of articles generally written by the researchers themselves. Level of reading is ordinarily not too difficult. Good source for maintaining some contact with a variety of fields.

Segal, E. M., and Lachman, R. Complex Behavior or Higher Mental Process: Is There a Paradigm Shift? *American Psychologist*, **27** (1972), 46–55. A discussion of a variety of causes of paradigm shifts.

Snow, C. P. *The Search*. New York: Charles Scribner's Sons, 1958. Originally published in 1934. A novel about scientists by an author who is intimately familiar with their world. They're just human.

Snow, C. P., *The Two Cultures and A Second Look*. New York: Cambridge University Press, 1963. The original essay examining the scientific and literary points of view, together with some later comments. An enjoyable classic.

Tullock, G. *The Organization of Inquiry*. Durham, N.C.: Duke University Press, 1966. Raises an interesting question: how do largely individualistic scientists manage to contribute to an endeavor that is essentially co-operative?

Vavorilis, A., and Colver, A. W. (eds.), *Science and Society*. San Francisco: Holden-Day, Inc., 1966. A collection of essays, many by eminent scientists, which attempts to outline the basic concepts and nature of the scientific enterprise.

Watson, J. D. *The Double Helix*. New York: Atheneum Publishers, 1968. A controversial and highly personal view of the formulation of theories regarding DNA. A sample of mixed emotions, attitudes, and values of some present-day scientists.

Weisskopf, V. F. The Significance of Science. *Science*, 176 (1972), 138–146. An interesting discussion of the plight of science; includes views both hostile and favorable to science.

White, A. D. *A History of the Warfare of Science and Theology*. New York: George Braziller, Inc., 1955. Originally published in 1895. Gives a lengthy and somewhat tedious recounting of the long and at times bloody struggle to separate science and dogmatic theology.

Woolf, H. (ed.), *Science as a Cultural Force*. Baltimore, Md.: The Johns Hopkins Press, 1964. Essays on the relations and problems between science and society. Brief but informative and important.

Index